Praise for *Out of the Blue*

"Friedman's book runs the gamut from superstitions held by ancient cultures about lightning, the history of lightning studies (including, of course, Ben Franklin), a ride-along with storm chaser David Hoadley, to human interest stories of lightning-strike victims...the real appeal of the book being the accounts of lightning-strike survivors, from Linda Cooper (struck by lightning four times in her life—no other known woman has been struck so many times) to those gathered at the Lightning Strike and Electric Shock Survivors International conference to the gripping rescue tale of mountain climbers hit by lightning at over 13,000 feet above sea level on Wyoming's Grand Teton."
—*USA Today*, "Weather Guys" blog

"John S. Friedman pans through time from ancient myths to scientists who are now delving through the mysteries which have surrounded this awesome and frightening subject. His greatest gift is painting a humanistic picture of a subject which has affected man since he began walking this earth." —Frank Field, TV weatherman

"Page-turner...For anyone cowering during a thunderstorm, the ideal reading material is [this] new book from John Friedman....The most moving parts of the book are [the] survival stories." —*Toronto Star*

"Journalist Friedman did his homework on this fascinating account of electricity from the sky—from its history in myth and fact to first-person recollections from people who were struck by lightning." —*The Sacramento Bee*

"Every outdoor enthusi
be enthralled by the
with an instant, unex
—*Sou*

D0813640

OUT *of the* BLUE

A History of Lightning:
Science, Superstition, and
Amazing Stories
of Survival

JOHN S. FRIEDMAN

DELTA TRADE PAPERBACKS

2009 Delta Trade Paperback Edition

Copyright © 2008 by John S. Friedman

Published in the United States by Delta, an imprint of The Random House
Publishing Group, a division of Random House, Inc., New York.

Delta is a registered trademark of Random House, Inc.,
and the colophon is a trademark of Random House, Inc

Originally published in hardcover in the United States by Delacorte Press,
an imprint of The Random House Publishing Group,
a division of Random House, Inc., in 2008.

Library of Congress Cataloging-in-Publication Data
Friedman, John S.
Out of the blue : a history of lightning : science, superstition, and amazing
stories of survival / John Friedman.
p. cm.
ISBN: 978-0-385-34116-5 (trade pbk.)
1. Lightning—History. 2. Survival. 3. Natural disasters. I. Title.

QC966 .F735 2008   2007050685
551.56/32 22

Printed in the United States of America

www.bantamdell.com

BVG   9   8   7   6   5   4   3   2   1

Text design by Carol Malcolm Russo
Cover design by Beverly Leung
Cover photo © Allan Davey / Masterfile
Back cover photo © Andrey Semenov / Shutterstock

*To Rathy, a different
kind of survivor,
and, as always, to Julia*

CONTENTS

ACKNOWLEDGMENTS

This book would not have been possible without the cooperation of lightning survivors and the families of those who did not survive. Like Coleridge's ancient mariner, some felt compelled to tell their stories; most told their experiences as a way to help others.

I owe special thanks to Joseph Drew for five decades of friendship and to Ariel Dorfman for a quarter century of friendship.

I also want to thank Michael Mierendorf, Richard Fox, Roger Salloch, Rosalyn Baxandall, Elizabeth Ewen, Patrick O'Sullivan, Isabelle Lemonnier, Barbara Solomon, Richard McCord, Henny Wenkart, Arnold and Judith Friedman, Louise Friedman, Roy Friedman, Sally and Harold Burman, Shakun Drew, Hugh MacKenzie, Cathy Lace, George Lombardi, James Llana, Brenda Levin, V. B. Price, Lois Stergiopoulos, Steve and Marie Roybal, Holly Stadtler, Ken Langford, and Melanie Aya-ay.

Special thanks to Steve Marshburn, Sr., Joyce Marshburn, Becky Loyet, and, above all, to the members of Lightning Strike and Electric Shock Survivors International, Inc. Valued assistance was provided by Jackie Skaggs and the Teton National Park Service rangers, particularly Chris Holm; the Lightning Data Center in Denver, Colorado; the Burn Center at Tampa General Hospital in Tampa, Florida; the National Weather Service; the Edsel Ford Memorial Library of the

Hotchkiss School; the New York Society Library; and all the members of the Department of American Studies/Media and Communications of the College at Old Westbury, State University of New York.

I greatly appreciate the ideas and editorial suggestions offered by my wife, Kathleen Friedman.

I am always grateful for the ongoing advice and guidance of Deirdre Mullane, agent extraordinaire of the Spieler Agency.

Finally, I want to express my deep appreciation to my editor, Danielle Perez, who is remarkably patient and always encouraging.

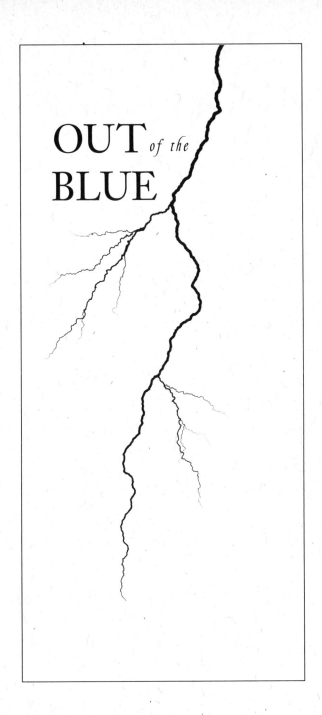

OUT *of the*
BLUE

AN
AWESOME
FLAME

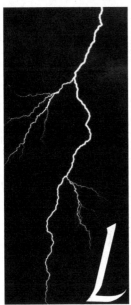

Everybody's afraid of lightning.
Maybe it's built into the genes.
It's a primal fear.
—DR. MARTIN UMAN

Lightning descends upon the American
landscape in fiery arcs across the Great Plains on lonely summer nights and in brilliant streaks over the Rocky Mountains
on lazy afternoons. It's also embedded in our oldest myths.

The stars are the campfires of the dead, and when we
die, the great Thunderbird, lightning flashing from its eyes,
carries our souls to the Milky Way. According to another
Native legend, the Sun, father of twin boys, gave them
magic arrows—lightning that strikes crooked and lightning
that strikes straight. One day, the twins heard rumbling like
the sound of an earthquake. It was the wake of the giant
Yeitso, who had smelled their scent. "How shall I kill them?"
the giant wondered. He fired four arrows at the boys, but
they missed.

Then the boy named Born of Water shot his own arrow
and hit Yeitso. And the boy named Monster Slayer shot his
arrow and it killed the giant. Afterward, the twins slayed

other monsters with their magical arrows, and they made a huge thunderstorm sweep across the land. When the storm ended, a place called the Grand Canyon existed where once other terrible creatures had lived.

Memories of the indiscriminate power and terrible fascination of lightning have remained with me since childhood. When I rode on horseback in the Sangre de Cristo Mountains in Northern New Mexico—the ancestral home of the Apaches, the Navajos, and the Pueblo Indians—thunderbolts flashed and crackled around me, and I feared I would never return home. The drama of the never-ending Southwestern sky and its storms—absent in the Northeast, where I have lived most of my life—shaped me in ways I am still exploring.

Ever since, I've wondered about the mysteries of lightning. What causes lightning, and what attracts and repels it? How can we protect against lightning, and when is it most dangerous? Why would one person walking in a field be struck and killed by lightning during a storm while his companion walks away unharmed? What happens physically to someone after being struck?

In his great novel *The Bridge of San Luis Rey*, Thornton Wilder observes that most "occasions of human woe had never been quite fit for scientific examination. They had lacked what our good savants were later to call *proper control.*" As he ponders the collapse of the Bridge of San Luis Rey, which killed several travelers in eighteenth-century Peru, Wilder's alter ego in the novel, Brother Juniper, collects "thousands of little facts and anecdotes and testimonies" to try to learn "why God had settled upon that person and upon that day for His demonstration of wisdom."

Lightning, too, was most often considered in earlier

periods of history to be a pure act of God, beyond scientific explanation. Today, the discoveries of science and medicine have altered our perspectives far beyond Brother Juniper's imaginings.

Still, being struck inevitably raises existential questions about life and death, destiny and divine retribution. What did I do to deserve this? What should I do now? After all, when lightning strikes, there is no human cause. Believing that the testimony of survivors would yield the "thousands of little facts and anecdotes" underlying the human dimension of lightning, I set out on a journey to record their stories. Their accounts reveal a remarkable blend of willful choice and random coincidence, science and superstition. They tell of heroism, pain, hope, and sacrifice. Above all, they tell of their own inspiring spiritual changes.

At least forty-four people were killed by lightning in the United States in 2007. The reported number is lower than the actual number because some deaths due to lightning are not recorded as such. Lightning is the second-leading cause of fatalities in the U.S. related to violent weather. It causes more deaths than earthquakes, tornadoes, or hurricanes. Only floods kill more people. But unlike these other natural disasters, lightning strikes are small, private tragedies, reserved for the unlucky few.

*L*ightning set my underclothes on fire," Roy Sullivan told a rapt audience watching the 1980s TV show *That's Incredible!* "Now, if you say that's not hot, I'd like to know what hot is."

A longtime ranger in Shenandoah National Park in the Blue Ridge Mountains of Virginia, Sullivan was born in Greene County, Virginia, on February 7, 1912. He was first hit by lightning in 1942, standing in a park lookout tower. He was lucky. His only injury was the loss of a big toenail. A brawny man with a broad, rugged face, Sullivan, who resembled the actor Gene Hackman, was struck again in 1969 while driving along a mountain road. This time the lightning only singed his eyebrows. But a year later, the outdoorsman was walking across his yard when lightning struck again, searing his left shoulder.

The fourth strike occurred in 1972, while Sullivan was working in a ranger station in Shenandoah National Park. It set his hair on fire, and he had to grab a bucket of water and pour it over his head to extinguish the flames. "I can be standing in a crowd of people, but it'll hit me," he said at the time. "I'm just allergic to lightning."

In 1973, while he was out on patrol in the park, Sullivan saw a storm cloud forming and drove away quickly. But the cloud, he said later, seemed to be following him. When he finally thought he had outrun it, he decided it was safe to leave his truck, but again he was struck. "I actually saw the bolt that hit me," he said. The next strike, the sixth, came in 1974 while he was checking a campsite near the Skyline Drive and left him with an injured ankle.

Then one Saturday morning in 1977, when he was fishing in a freshwater pool, lightning struck Sullivan for the seventh time—hitting the top of his head and traveling down his right side. With his hair singed and burns on his chest and stomach, he hurried to his car. But still he kept his wits about him. He later told a reporter that as he stumbled back down the trail, a bear appeared and tried to steal three trout from his fishing line. But Sullivan had the strength and courage to strike the bear with a branch. He recalled that it was the twenty-second bear he had hit on the head during his lifetime.

Sullivan owns a place in the *Guinness World Records*, not for the number of times he's decked a bear but for the distinction of being struck by lightning more recorded times than any other human being. Some reports state that he was hit an eighth time in the early 1980s. "Naturally people avoided me," he once recalled. "For instance, I was walking with the chief ranger one day when lightning struck way off. The chief said, 'I'll see you later.'"

On the one hand, Sullivan seemed to attract lightning. (He was dubbed "the human lightning rod" by the media.) On the other hand, he appeared to have some natural physical defense against its effects—despite the number of times he was struck, he wasn't killed or even seriously injured. A member of the Shenandoah Heights Baptist Church, Sullivan had conflicting thoughts about his own fate. He believed that an unseen force was trying to destroy him, and he became convinced after the fourth strike that the next bolt would kill him. Still, he once told a reporter, "I don't believe God is after me. If He were, the first bolt would have been enough."

After supposedly being rejected by the woman he loved, or perhaps from the fear and dread of future strikes, Roy Sullivan shot and killed himself in 1983 at the age of seventy-one. He was living at the time in a town called Dooms.

*L*inda Cooper seems to have it all. She is happily married and lives in an upscale neighborhood of Spartanburg, South Carolina. She has three daughters and three grandchildren.

She likes her job as a computer lab supervisor at the local elementary school and generally enjoys life.

And yet, she suffers. Linda Cooper has been hit by lightning four times in her life. No other woman, as far as is known, has been struck as many times. Men account for about four times more lightning fatalities and injuries than women, as men are more likely to engage in agriculture, construction, and recreational outdoor activities.

I am sitting on a bench with Linda Cooper in a hallway of the MainStay Suites in Pigeon Forge, Tennessee, where a conference for lightning survivors is taking place. She is wearing a tailored red jacket, a blue dress, and a silver necklace. She speaks with a slight Southern accent and is poised and attractive. Complimented that at fifty-seven she looks ten years younger than her age, she replies brightly, "Makeup and curlers do wonders."

Cooper was born in Atlanta in 1950 and grew up in Miami. She was first struck by lightning on September 15, 1983, which had been a typical September day in Ft. Lauderdale, Florida—dismal and gray. It had been raining on and off all morning, and just after one p.m. it started to sprinkle again.

Setting out on a round of errands, Cooper had parked her car in front of the Coral Ridge post office, where she was going to mail a package. When she stepped onto the sidewalk, "it was like a hand grenade going off in my face," she recalls. "All I remember is a blinding white light and the loudest sound I have ever heard or could ever imagine hearing."

The next thing she knew, she was standing up, brushing off her dress, and wondering why she was all wet. Confused and in shock, she walked into the post office. On her way in, she turned and saw a man in his car staring at her. "On his face was a look of horror." They never spoke and he drove away quickly. But to this day, Cooper wonders what he saw.

She went up to the counter to mail her package and told the clerk that she had just been hit by lightning. "We know," the clerk replied. "It shook the building." Lightning had struck a flagpole and walloped Cooper on the left side of her head. .

Cooper then returned to her car and drove to her aerobics class. (Remarkably, shock victims often continue their daily routines immediately after being struck.) Afterward, she telephoned an HMO clinic and described what had happened. Since she seemed to have recovered from the shock, the doctor told her that she would be all right.

Feeling "weird," she nevertheless went the next day to the preschool where she worked. After classes had ended, a parent came into the office and stood close by. Looking puzzled, she said, "I feel electricity but I can't tell where it's coming from." Cooper crossed to the other side of the room and asked if the woman still felt it. "No," she answered, "it's coming from you."

Still feeling peculiar, Cooper went to a doctor a few days later, who told her that she looked fine and that she had just bruised herself. When Cooper said that her leg hurt, the doctor replied, "You probably twisted your ankle; try to stay off it."

Three weeks passed and Cooper still felt unwell. She couldn't keep her eyes open, couldn't get out of bed. "I was sick as a dog and all I wanted to do was die." She finally went to the hospital, where, she recalls, "they freaked out.

"My back had turned blue, black, purple-green. They

started doing every kind of test imaginable and they kept me in the hospital for six days, after which they said, 'You're lucky to be alive,' and sent me home."

For nine months, she suffered massive headaches and terrible pain. Finally she saw a neurologist who prescribed phenerol, a mild narcotic, which she took for a little while. But it didn't help relieve her feelings of being isolated and alone, a common complaint among lightning survivors.

"If I had a cast on my arm or leg, people would have had compassion, but seeing no outward signs of anything wrong, they were a bit tired of my situation."

Cooper tried to resume her normal life, but things had changed dramatically. She couldn't remember how to add and subtract. (Again, a common occurrence.) When she went shopping, a friend had to accompany her to complete her sentences. She and her husband divorced and she changed jobs. The first year after the strike was "a year of hell." Physically, mentally, emotionally, she was exhausted and in pain.

Some years later her daughter telephoned and said, "Mama, they want you to be on *The Oprah Winfrey Show* because they are looking for people who have survived terrible accidents, and you need to do it because it's all about you." Cooper hesitated at first but finally agreed. On the show she met a survivor named Edwin Robinson, who in 1980 had gone into his backyard during a thunderstorm, when lightning ricocheted off a tree and hit him. He claimed that when he regained consciousness the bolt had improved his vision and his hearing. He told her about an organization called at the time Lightning Strike and Electric Shock Victims International, which assisted people who had suffered from lightning injuries. She contacted the group and was helped tremendously by talking with other survivors: "It's my life-line to the real world that I live in, which is different from the real world that you live in."

Over the years, her condition improved slowly. She went back to school to learn how to do simple arithmetic. She started doing aerobics again and lifting weights. She re-married. After the eight years that it took for her to pass the two-year course to become a certified paralegal, she received her certificate on May 5, 1993. Life was returning to normal.

Three weeks later, lightning struck again.

Cooper and her family were now living in a town house by the ocean in Hillsboro Beach, Florida. She was in the kitchen, talking on the telephone. The sky was clear. Suddenly there was a loud boom and the charge smacked her in the face. For the next few months she experienced nausea and felt as though she had the flu, but this time she didn't go to a hospital: "I knew that I was breathing. I knew that my heart hadn't stopped, and I knew that they wouldn't do anything for me."

Soon after, her husband, Gordon, received an offer to join a law firm in Spartanburg, his hometown. There, she took a part-time job as an executive assistant to the president of a local corporation.

On July 11, 1994, Cooper was home alone making Jell-O for dessert. It had been raining earlier but the storm had passed. Reaching to turn on the tap, Cooper heard "a loud POP—like a flashbulb in the old-time press cameras." Lightning had struck nearby, traveled through the plumbing lines, through the faucets, and up both her arms. "It felt like somebody took a match and lit my arms. I felt like I was on fire." She hurried to the freezer and leaned over it until her body had cooled down.

She shows me the scars from the burns on her arms and legs. When she called her doctor to tell him about the incident, he prescribed Zoloft, an antidepressant, and Ultram, a pain reliever she still takes.

The fourth time, Cooper was sitting in a car with a friend in the parking lot of the Hillcrest Mall in Spartanburg on August 22, 2003, waiting for a storm to pass. Her friend was using a cell phone and handed it to Cooper just as lightning hit. The current passed through her friend's body, destroying the cell phone, and into Cooper's thumb. Her friend took the brunt of it but suffered only minor injuries. Luckily, Cooper received only a small burn on her hand.

Cooper describes the four strikes to me. The first was like being hit by "a Mack truck"; the second like being hit "by a bike"; the third by "a truck"; and the fourth like "falling off a bike." "Each shock exacerbated the others," she says. "But the one that hurt me the most was the first. When I get tired my speech slurs, my left foot goes out on me." In addition, Cooper's car has been struck twice by lightning, her garage a couple of times, and she has lost telephone service in her house to lightning at least three or four times.

Cooper, who was brought up as a Southern Baptist but converted to Presbyterianism in her twenties, doesn't believe that God punishes anyone. She's been influenced by the book *When Bad Things Happen to Good People* by Harold S. Kushner, which she read after she was struck the first time. "I believe that God gives you the strength to endure whatever happens while you are living on earth." She gives an example of her own attitude: "If you hear an ambulance, don't pray, 'Oh, my God, oh, my God, don't let it be my house.' What you should pray is, 'Oh, my God, if it is my house, let me have the strength to get through it.'"

As a result of the lightning strikes, her faith has become stronger. "I believe in God because if I didn't believe in God I wouldn't still be here. I would have taken my life a long time ago."

She adds, "You know life is a circle. It's not going to go away when I die. I can only put out a little ripple, but where my ripple goes I want to leave a good impression."

At a recent meeting of the lightning survivors' organization, a woman whose husband was suffering from a lightning strike met Cooper in the ladies' room and fell into her arms, sobbing. "I know why you lived," the woman tells her. "To give testimony to people like us. To give us hope."

I'm in the coffee shop of a motel in Oklahoma City on a warm May afternoon in 2007 with Carl Mize, who is sipping iced tea. He is muscular and solid from years of competing in high school as a wrestler and then as a rodeo bull rider. He is wearing sandals, Bermuda-length blue-jean shorts, and a red and gray T-shirt that reads "Oklahoma Sooners." The waitress greets Mize: "How ya doin', Sparky?" He is used to such teasing, as he occupies a unique role on the planet: As best as can be determined, he has been struck by lightning more times than any other living person.

Mize was born in Sulphur, Oklahoma, on the first day of spring in 1960. When he was about thirteen years old he began competing in rodeos. "I've been hooked a few times," he said, "and broke ribs when a bull fell with me against a fence. My body hurts, but I don't know if it's from that or from the lightning."

He recalls the first time he was struck, in the summer of 1978 in Claremore, Oklahoma, the year he graduated from high school. "I was in the bull-riding competition when there came a big rainstorm and they got the rodeo over—except for the bull riding," he says in his thick Oklahoma accent. "It was thundering and lightning and carrying on. I picked up my bull bag and ran to my truck. When I grabbed the door handle, lightning struck and knocked me on my rear end. I got my butt all mashed. It made me real sore. They rescheduled the rodeo for Sunday, and I was there—riding."

He seemed not to have suffered any serious medical problems, but later he was at work when "I kinda went out. I had a high fever. I woke up the next day in the hospital. It was like I had Bell's palsy," a condition that causes the facial muscles to weaken or become paralyzed. He had to stop working for a month.

Mize says he had forgotten about the second time he was struck by lightning until a buddy recently reminded him. It was about 1985 or 1986. He was helping move a child's outdoor playhouse on a flatbed trailer when lightning struck a transformer above his head, shooting out flames. The bolt knocked a crowbar out of his hands. "I felt I was shot by a twenty-two or something."

The third time he was hit was on August 9, 1996, while Mize was working as a utility man at the University of

Oklahoma in Norman. He and his coworkers were responsible for maintaining all the outdoor electricity on the campus—fixing lights, locating underground cable at construction sites, and helping with water, sewer, and gas problems. "My job was outside. Rain, sleet, shine, cold. I'd rather do that than be cooped up inside a building."

He was reaching down into a hole, trying to put a cable together, when lightning hit a forty-foot bald-cypress tree nearby, split it, bounced over to a streetlight pole, zapped through his hand, and knocked him unconscious. When he woke up, he told his foreman, "I think somebody just hit me with a baseball bat." He had a large burn mark on his chest and his whole body was sore. He remained in the hospital for three days and underwent a battery of tests, but no serious injuries were found.

The fourth time he was hit was on May 3, 1999, working with his animals—he raises sheep, goats, and pigs. His wife told him there was a tornado in nearby Chickasaw and that she wanted to go to their son-in-law's parents' house, where there was a storm cellar. Mize refused. Twenty minutes later she came out of the house again and said, "The thing is still on the ground and we need to go." Still Mize refused. She came out a third time and said, "You stay your stubborn ass here. Me and the kids are going."

Mize relented, but while everyone else huddled in the cellar, he still stood "underneath a tree like a stupid person," though he knew better. A menacing green and black cloud hung overhead. Mize was holding the chain of a swing, watching the cloud, when lightning struck the tree, knocked the top off, and zapped him to within four or five feet of the house. It made him "more or less sore on one side," but he didn't go to the hospital.

On August 5, 2005, Mize was working to help repair a broken water main on the University of Oklahoma campus. All the members of the repair crew were wearing rubber boots except Mize, who was wearing tennis shoes with holes in them. A storm was approaching. "One of the guys said, 'I don't want to be standing next to Carl. He's been struck too many times.' We all laughed about it. I looked at a guy next to me and said, 'I can't believe he's going to walk away.' I tell them the best place to be is beside me. It's going to hit me, not you.

"The next thing I knew lightning hit me."

Apparently, the bolt hit the backhoe, then dispersed through the water and wet grass and finally through Mize's foot. His foot bounced off the ground, knocking him backward. The lightning never touched anybody else. Mize was taken to the hospital, where he remained for four days. There were problems with his heart rate. He had high blood pressure and his foot was numb. But he soon returned to his job.

After Mize was hit the fifth time, CBS, ABC, and FOX all wanted to interview him. A producer from CBS's *The Early Show* telephoned and offered to have a car pick him up in an hour to take him to a motel so he could be interviewed early the next morning in a studio. "I told her, 'Ma'am, I have a lot to do tonight. I have to attend a dinner and take care of my animals.'" ABC was willing to fly him to New York that night and actually sent someone to his house. But he had given his word to CBS. He appeared on *The Early*

Show and was also interviewed by reporters from Russia, Europe, and across the United States. Fuji Television in Japan reenacted scenes from the lightning strikes for a documentary.

Interviewers couldn't help but ask the obvious. "What the heck were you doing out there and wearing tennis shoes with holes in them instead of rubber boots?"

"I'm not scared of lightning," Mize replies. "I always think it can't happen again. Maybe that's the wrong way to think. Maybe I'm stupid. I'm not scared of dying."

The sixth strike came in August 2006. A storm was approaching as Mize was putting a tarp on top of a stack of alfalfa hay. He picked up an old tire to weigh down the tarp, and as he was throwing the tire on the stack, blinding light slashed through the air, throwing him to the ground.

"The first thing I thought—it might have knocked me out—was I'm not dead, and there is no way this could be happening again. It rained a little bit. I was muddy. I went in the house and took off my clothes. My wife came in and asked what's wrong. I said, 'I've been hit.'"

" 'Did that ram hit you?' she asked.

" 'No, lightning did.' "

Mize has also had an ongoing dangerous relationship with electricity. His house burned down about fourteen years ago in a fire caused by an electrical short. And he's been electrocuted a number of times. "I was never hurt," he recalls, "but it scares me more than anything." He remembers in particular the time he went to fix a fountain in a duck pond at the university. He was in a wetsuit floating in the water. The electrician had told him that the current to the pond lights was turned off. But when Mize cut a wire under-water, the water fizzled and smoke rose above the surface. "It was four hundred eighty volts. Enough to kill someone. I never did understand it."

In addition, Mize has had sheep, a horse, and a goldfish killed by lightning. The goldfish was swimming in a water tank outside his house when lightning hit a nearby tree.

He takes a long pause. "Sometimes when I'm driving from here to there, I turn the radio off and I reflect on life and often think, why?

"My mom is really religious and always says there must be some reason why I've been struck so many times. I'm a Baptist but I don't go to church regularly. But I tell her God must like me pretty well—I'm alive. Some people say, 'Man, you're unlucky.' I say, 'No, I'm lucky.' There's a fine line be-tween being unlucky and lucky.

"I remember in grade school two friends of mine were fishing when a storm came. There was a tree in the middle of a pasture. They ran and got under it. Lightning struck the tree. It killed one boy and not the other." Mize shakes his head. "Why?"

*A*s I set out, I sought to understand the long human ex-perience of lightning. From the ancient myths that primitive cultures created to explain lightning's spectacular force, to Benjamin Franklin's renowned—and probably foolhardy—experiments during a thunderstorm, to modern, high-tech testing grounds where scientists are still teasing out light-ning's secrets, I would examine our changing concepts of this powerful natural force. I would look at where and when lightning is most likely to occur and what precautions golfers, hikers, and others should take to avoid disaster. With storm tracker David Hoadley, I would chase lightning and thunderstorms across the great Midwest, and I would speak with the survivors of the most heroic lightning rescue in American mountaineering history. Wondering about the devastating effects of a lightning strike on the body, I would find that the problems survivors face are often baffling to those doctors who are not among the few specialists in the field. And I would listen to survivors recall their extraordinary near-death and out-of-body experiences, and their painful steps to recovery, and tell how lightning has profoundly affected their personal philosophies. I interviewed in-depth and in-person about twenty-five survivors, and sometimes their families; spoke at length on the phone with at least ten others; and read the accounts of about a hundred more.

Most survivors undergo life-altering transformations. "The day my life changed forever" is how they describe it. What I would learn from their stories touches upon spiritual and psychological experiences; suggests safety measures that could save lives; and, finally, reveals that encounters with thunderbolts may be more about the future than the past. Like the Native Americans, the survivors take lightning into their lives, and, often unwillingly, make it part of their being.

In the end, this book is the story—historical, scientific, meteorological, cultural, medical, psychological, and spiritual—of a blind and ungovernable force that is indifferent to suffering and seems not to distinguish between the rich and the poor, the young and the old, the strong and the weak. It is an unconventional history, from the cavemen to the

X-Men, of a spectacular weather phenomenon that may become only more common as global warming intensifies. It is a tale of tragedy and survival, stretching from the Tetons to the Central Plains to the storm-plagued coasts of Florida. Above all, it is a story of the human response through the ages to what novelist Herman Melville called "God's burning finger."

THE MOUNTAIN CLIMBER:

THE ASCENT

Is there not an appointed time to man upon earth?

—JOB 7:1

The July 2003 morning began with cool temperatures in the fifties, as the thirteen climbers began their ascent of the Grand Teton, the highest peak in the majestic Teton Range. By noon it was seventy-three degrees even in the low valleys. By the early afternoon, the warm Wyoming air was moving up the mountain, past the Jenny Lake Ranger Station and the climbers who were slowly ascending the Exum Ridge route along the 13,770-foot spine of "the Grand." Typical of a July afternoon in the Tetons, a menacing mass of black storm clouds had drifted toward the mountaintop. The climbers, who were only 700 feet from the summit, looked at the sky and knew that the clouds were an ominous sign, warning them to think twice before continuing on.

As the sunny morning gave way to afternoon rain, the walls of the Grand became treacherous. It's hard enough finding a place to grip the smooth walls when they are dry but nearly impossible after a rain, when they are deceptively slippery.

Rod Liberal is halfway up the rock face when he hears an eerie buzz, as if swarms of bees are gathering on the mountain. The hair on his arms stands up, and for a brief moment an instinct courses through his body, warning him that something terrible is about to happen. Then out of nowhere flares a burst of heat and light and a deafening noise, making it seem that the very mountain itself is trembling. The bolt of lightning scorches the air, striking the narrow ledge above him.

Suddenly, everything is black and Rod is spinning around, falling. He doesn't know what's happened. He only knows he's going to die.

Rod's friend Clinton Summers has been belaying him up a hundred-foot wall and had attached Rod's rope through the carabiner clip seconds before the bolt hit, but that doesn't prevent Rod from plunging—five...ten...fifteen feet. He bounces off the smooth rock, falling until the slack plays out, jerking him up abruptly. He is hanging from his harness in an upside-down position, bent backward, belly facing the sky, 13,000 feet in the air.

The first thing he feels is something rough in his mouth, like sand. He tries to straighten himself out, but his left arm is completely dead. His right leg is also lifeless. His back hurts. It's probably broken. The pain is terrible. Opening his eyes, he stares at an empty sky. He tries to scream, but the sound echoes only in his mind.

He can't see anybody. He can't hear anybody. Like the castaway Pip in *Moby Dick*, abandoned in the awful lonesomeness of the ocean, he floats in heartless emptiness.

*B*orn in Brazil in 1975, Rodrigo Liberal came to the United States with his mother when he was fifteen years old. Moving from New York City to Florida, he worked various jobs at delis, restaurants, and construction sites, and delivered newspapers in South Beach. Then, a few years later, he found a job working at night with American Express, where he trained to become a software developer.

Thin, wiry, and fit, with eyes like shards of coal, he enjoyed playing sports and was particularly passionate about hockey. At an ice-skating rink in Florida, he met Jody Lemanski and married her in 1998. Four years later, they moved from Salt Lake City to Idaho Falls, Idaho, where he had found a job as a programmer. "By the time we got to Idaho Falls I was pretty into climbing," he remembers. "That was probably my main activity."

At work, Rod had become friendly with colleagues he describes as "big-time climbers." In the early summer of 2003, he joined them on a climb in the shadow of the Grand Teton Range. They told him that for a year they had been planning to scale the Grand, and they invited him along on the trip two weeks later. "I was a little hesitant because I had never done anything like that," he said. "But I also knew it was the chance of a lifetime." Most of the climb would be spent hiking, and Rod was in great physical condition.

"I was thrilled. It was the Grand Teton—one of the top fifty climbs in the country. It was late July, a perfect time of year. We took Friday off from work and started hiking up the mountain to a place called the Meadows. It pretty much took us all day to get there. It was a beautiful day. A perfect hike. We arrived near sundown." The climbers cooked dinner and pitched their tents. They needed the rest. They were each packing about forty or fifty pounds—food, water, and all their climbing gear. One of the climbers slept out under the stars. It rained that night but he didn't mind.

They encountered other climbers and hikers along the popular route. Rod and his friends talked for a while with a young woman, a geology student, who pointed out to them a visible trace of dark rock exposed by erosion, called the Black Dike, stretching down the mountain to the valley below. It formed after molten rock had poured into a crevice in older gneiss rock before the Tetons were pushed up by tectonic forces.

"She asked what we were doing the next day, and she said she was either going to take one trail or another depending on the weather. She mentioned something about lightning and said thunderstorms were predicted. I mentioned this to my friend Jake, and we kind of looked at each other but didn't make much of it. It's kinda like an earthquake. You think, what are the odds?" Actually, the odds of getting hit by lightning during recent years in the United States are about one in 750,000, and the odds of dying from lightning in the United States are slightly less than one in six million annually.

At 3:46 p.m., a cell-phone call comes in to the ranger dispatch center in Moose, Wyoming, from a climber on the mountain. It is only minutes after the lightning strike. The usually unflappable dispatcher tenses up and quickly sends out an emergency call: "Five people struck by lightning; one person not breathing; one person hanging upside down and breathing; three people missing and not responding."

Ranger Brandon Torres, rescue coordinator for the day, is in his office at the Moose Visitor Center, speaking on the phone with a woman who can't find her twelve-year-old son after the boy had gone into the woods to relieve himself and gotten lost. His radio squawks: "Four-three-one. Teton Dispatch." Juggling the phone, Torres picks up the radio, and the dispatcher gives him the grim news.

Torres delicately ends the conversation with the frantic woman. He glances out the window. It is raining. An experienced climber and mountain rescuer, the thirty-six-year-old Torres refers to himself as a "West Coast kid" who was "really freaked out by all the lightning" when he first got to the Tetons. Now he has to act quickly. He has only a few hours before dark, when a rescue will be impossible.

He immediately asks the dispatcher to order a helicopter and page all rangers in the Jenny Lake area, which is in the heart of the Tetons.

Torres races in his Chevy Suburban to the rescue cache, a cabin in Lupine Meadows. On the way, he asks the dispatcher to transfer the climber's cell-phone call to his car phone. He tells the climber, Bob Thomas, who has flown from California to join the climb, to make sure everyone is tied to anchors and to determine the location of the injured climbers and the extent of their injuries. Thomas tells him that he is trying to contact everyone using the portable radios that the climbers are carrying. Torres says that he will call him back in ten minutes, after briefing the rangers.

When Torres arrives at the cabin, the other rangers are

getting their equipment ready. A dozen more rangers are on duty, but most are dispersed across 220 miles of hiking and climbing trails. Inside the cabin, Torres briefs the rangers, writing on a chalkboard the names of the climbers, their approximate locations, and sketchy details of their injuries provided by Thomas. He tries to call Thomas back but gets only a voice-mailbox recording.

Torres briefly recalls that a month before a twenty-three-year-old woman had fallen to her death in Garnet Canyon. Prior to this rescue effort, only one person had ever been killed by lightning in Grand Teton National Park.

Setting the group in motion, Torres has no idea that this will be the most complex recovery operation the Teton rangers have ever attempted and the most difficult and boldest lightning rescue in American mountaineering history.

THE ANGRY SKY

The thunder,
Winged with red lightning
and impetuous rage
—JOHN MILTON, *PARADISE LOST*

To the earliest humans, lightning was a magic fire from the sky. Unlike other violent weather such as tornadoes or hurricanes, lightning is creative as well as destructive. Its fire is to be feared, but if tamed it can keep humans warm and dangerous animals at bay, and it often brings with it life-giving rain. It is little wonder, then, that almost as soon as humans began to turn the daily struggles of their lives into Stone Age art, cavemen depicted lightning.

The oldest representation of lightning may be an image found in the Kimberley District in remote western Australia, in prehistoric rock engravings estimated to be twenty thousand years old. In the portrait, a human figure, perhaps a member of a lightning cult, is shown with a mass of unruly hair, as if animated by static electricity from the clouds, resembling the sensation one feels during a lightning storm that one's hair is about to stand on end.

The depiction of lightning in images and stories has been seen in almost all cultures throughout human history. Lightning is usually understood as an emblem of power, change, providence, and destiny or as the embodiment of piercing light that reveals divinity. In ancient Egypt, a deity named Typhon was believed to hurl lightning, while a seal from Mesopotamia from about 2200 B.C. shows a deity standing on a winged creature holding a bundle of thunderbolts in each hand. The Babylonians envisioned Ishkur, a deity who with one arm carried a boomerang for making thunder and with the other hoisted a spear to hurl down lightning. The Chinese told of Lei Tsu, the god of thunder, who wore a halo of fire and made thunder by beating drums; the ancient Vedic books of India described Indra, who carried thunderbolts on his chariot; and some statues of Buddha show him holding a thunderbolt with prongs at each end. The Yoruba of western Nigeria revered Shango, who created thunder and lightning by casting "thunderstones" down to earth, and wherever lightning struck, priests searched the surrounding area for the divine rocks. The Mayans worshipped Ah-Peku, the Irish Tuireann, the Polynesians Haikili, the Basques Orko, and the Slavs Perun. For the Tibetans, lightning is associated with Vajra—both phallus and sword. Fire represents the male force, the yang, powerful and phallic.

Like the Native Americans, who venerated the great Thunderbird—lightning issuing from its eyes, thunder beating from its wings—the Bantu tribe in Africa believed that lightning was produced by a magical creature called Umpundulo, whose bright plumage and powerful wings created lightning and thunder. Even in relatively recent times, Bantu witch doctors went out into storms with loud shouts and warnings blown on their flutes to entreat the Thunderbirds to fly away.

But perhaps nowhere is the idea of divine power as closely identified with the phenomenon of lightning as in the stories of the Greeks' most powerful god, Zeus. As early as 1000 B.C., tribes in Greece were worshipping a weather deity who would evolve into the figure of Zeus, and by about the eighth century B.C., when the historian Hesiod wrote his *Theogony*, which became for the Greeks a definitive source about the origins of the deities and the creation of the universe, the myth of Zeus had taken shape. According to Hesiod, the Cyclopes, giants with one eye in the middle of their foreheads, had given Zeus lightning, which became his favored weapon in his battles with his brothers and sisters, the Titans, and with his father, Cronus, in an epic struggle to control the universe.

As he waged war, Zeus "came forthwith, hurling his lightning: the bolts flew thick and fast from his strong hand together with thunder and lightning, whirling an awesome flame." After the Titans were conquered, Zeus was challenged by Typhon, "a monster with a hundred heads." Again relying on thunder and lightning, "the bolt that never sleeps," Zeus destroyed Typhon. Finally victorious, Zeus reigned in heaven, "holding the lightning and glowing thunderbolt," while the Cyclopes—Arges (thunderbolts), Steropes (lightning), and Brontes (thunder)—served as his blacksmiths.

Again and again throughout Greek literature, Zeus deploys lightning as a way of enforcing divine retribution to kill those who have incurred his wrath. For example, in *The Odyssey*, Homer tells of the arrival of Odysseus and his crew on the island of Apollo's sacred cattle, where they are warned not to eat the animals. Driven by hunger, the crew ignores the warning. Jove (Zeus) is furious, and as soon as their ship sets sail, they are met by a fierce storm in which "Jove let fly with his thunderbolts, and the ship went round and round, and was filled with fire and brimstone as the lightning struck it." All on board perish except Odysseus.

Countless others in Greek mythology die from lightning strikes caused by Zeus's anger, most often for sexual transgressions or an arrogant grasping at power. When Zeus discovers Iasion, a mortal lover, copulating with the earth deity Demeter in a field, he strikes Iasion with a lightning bolt, and the unfortunate youth is then said to have been placed among the stars in the constellation Gemini. Likewise, when Anchises boasts of being Aphrodite's lover, he is hit by lightning.

Lightning strikes were especially instructive when used against those who dared assume the powers of a god. Asclepius, the deity of medicine and healing, was struck by a lightning bolt after Zeus feared that Asclepius's ability to restore life would make humans immortal. And when Phaeton convinced his father, Helios, the sun deity, to allow him to drive the chariot of the sun across the sky, he came so close to the earth that he almost set it on fire, and Zeus immediately killed him with lightning. The Greek historian Herodotus sums up Zeus's methods when he writes in *The Histories*: "You may have observed how the thunder-bolt of Heaven chastises the insolence of the more enormous animals, whilst it passes over without injury the weak and the insignificant: before these weapons of the gods you must have seen how the proudest palaces and the loftiest trees fall and perish...."

Despite such disastrous precedents, a number of ancient thinkers were not afraid to challenge the popular conception of lightning as the embodiment of divine power. In his satire *The Clouds*, the Greek playwright Aristophanes mocks the notion of lightning with bawdy humor as Strepsiades, an old tightwad, asks Socrates what causes thunder. Socrates replies that after overeating you get "a stomachache and then suddenly your belly resounds with prolonged rumbling." Strepsiades agrees: "I get colic, then the stew sets to rumbling like thunder and finally bursts forth with a terrific noise. At first, it's but a little gurgling pappax, pappax! Then it increases, papapappax! And when I take my crap, why, it's thunder indeed."

Socrates: "Well, then, reflect what a noise is produced by your belly, which is but small. Shall not the air, which is boundless, produce these mighty claps of thunder?"

Strepsiades: "And this is why the names are so much alike: crap and clap. But tell me this. Whence comes the lightning...?"

Socrates: "When a dry wind ascends to the Clouds and gets shut into them, it blows them out like a bladder; finally, being too confined, it bursts them, escapes with a fierce violence and a roar to flash into flame by reason of its own impetuosity."

In the Roman Empire, where lightning became even more closely aligned with the gods than among the Greeks, to challenge lightning and thunder was to mock the deities and risk their punishment. According to Roman legend, Amulius Silvius, one of the early kings of Alba, in Northern Italy, claimed to be the equal of or superior to Jupiter. To support his pretensions and overawe his subjects, he constructed machines that mimicked the clap of thunder and the flash of lightning, but he paid for his impiety, perishing when his house was hit by a thunderbolt.

Similarly, another Roman king, named Salmoneus, a founder of the city of Salmonia, created mock thunder by dragging bronze kettles behind his chariot while hurling blazing torches in imitation of lightning. His ersatz thunder and lightning were not merely noisy exhibitions meant to impress the populace but were enchantments designed to bring about rain and storms. He, too, was killed by a thunderbolt.

In ancient Rome, places where lightning had struck were considered holy and became the sites of oracles and temples, consecrated to Zeus, "who came down in the flash from heaven." In the first century B.C., Caesar Augustus, who always carried a sealskin with him for protection against lightning and took refuge in an underground room at the first sign of violent storm, erected an altar to Jupiter at the place where he once had a narrow escape from lightning.

From their forebears the Etruscans, who lived on the thundery slopes of the Alps where lightning was a more familiar occurrence than in the rest of Europe, the Romans inherited the practice of lightning divination—trying to determine divine will by signs and portents in the sky. The Etruscans codified their beliefs in the Books of Lightning, or *Libri Fulgurales*, a sacred text outlining the art of ceraunomancy (divination by observing lightning), which became part of Roman statecraft and lasted for many centuries.

In the Roman Empire, divination was entrusted to the College of Augurs, which was formed as early as 300 B.C. and continued until the fourth century A.D. Thought to be able to discern the will of Jupiter, the augurs had immense power. Their reports became politically useful tools, manipulated to postpone unwanted meetings, delay the passage of laws, or prevent the election of certain magistrates. An augur tried to divine the will of the gods primarily by observing the southern sky for random signs of lightning, birds, and shooting stars. A lightning bolt passing from left to right was a favorable omen; a lightning bolt passing from right to left was a sign that Jupiter disapproved of current political events. Whenever the augurs reported evidence of lightning, the magistrates of Rome were required to cancel all public assemblies on the following day. An augur's promise of "I will watch the sky" became a euphemism for impending political upheaval, notes historian Philip Dray.

When a new consul prayed for favorable omens before he was to take high office, it was a matter of custom and political expediency for the augurs to report that lightning had been seen, coming, of course, from the left side of the sky. The Roman historian Suetonius mentions in *The Twelve Caesars* that in the eight months preceding the assassination of the emperor Domitian in 96 A.D., so much lightning had been recorded that Domitian cried out, "Now let him strike whomever he pleases." On the day he was assassinated by his own steward, Domitian had ordered the execution of a German soothsayer who was charged with claiming that lightning portended a change of government.

*A*fter the ancients, the idea that lightning was a sign of divine anger and retribution lingered, but lightning also came to be seen as a more benign symbol not only of God's power but also of His visible presence and involvement in human affairs. In the Bible, the fear and dread of lightning shared by the Greeks and Romans is tempered. Lightning is still described as a force that strikes terror into those who behold it, but it is not commingled with divine punishment. When God appears to Moses in Exodus (19:16), "there were thunders and lightnings...so that all the people that *was* in the camp trembled." In Samuel II (22:15), David praises God for delivering him from his enemies—"And he sent out arrows, and scattered them; lightning, and discomfited them"—and in Job (38:35), Job asks God for a sign of revelation: "Canst thou send lightnings, that they may go, and say unto thee, Here we *are*?"

In the New Testament, lightning is often a revelatory sign, an extraordinary celestial omen likened to the coming of Christ. In Matthew (24:27), for example: "For as the lightning cometh out of the east, and shineth even unto the west, so shall also the coming of the Son of man be." In Luke (17:24), Jesus tells His disciples: "For as the lightning, that lighteneth out of the one *part* under heaven, shineth unto the other *part* under heaven; so shall also the Son of man be in his day."

*F*or centuries, Church authorities challenged an enduring custom from Roman times that associated protection from lightning with certain living plants. The laurel bush was believed by Roman emperors to provide protection from lightning. "Whenever the sky wore an ugly look," the Emperor Tiberius would put on "a laurel wreath which, he supposed, would make him lightning-proof," according to Suetonius's *The Twelve Caesars.* But it was a belief in the mystical connection between lightning and the oak tree that became especially widespread, a belief that has survived in its way until the present.

In Rome, Jupiter's great temple on the Capitoline Hill was said to have been built by Romulus beside an oak, which became Jupiter's sacred tree. In Greece, the Oracle of Dodona, the oldest of the Greek oracles and one of the most famous sanctuaries in the ancient world, was located in the northwest mountains, where Zeus was worshipped after lightning had struck an oak tree there. "There was a tradition in the temple of Dodona that oaks first gave prophetic utterances. The men of old, unlike in their simplicity to young philosophy, deemed that if they heard the truth even from 'oak or rock,' it was enough for them," Socrates says in the *Phaedrus.* But in 391 A.D. the Oracle at Dodona was destroyed when the revered oak tree was cut down on the orders of Roman Emperor Theodosius, a proponent of Christianity who wanted to stamp out all vestiges of pagan worship. Archaeologists have since replanted oak trees at Dodona, where visitors can sit beneath their branches and listen to the whispers of the past.

In the Middle Ages, oak trees were often cut down in the effort to convert pagans to Christianity. In the eighth century A.D., to demonstrate how utterly powerless were the gods in whom pagan believers placed their confidence, St. Boniface ordered destroyed an ancient oak standing in Geismar, Germany, that was said to be sacred to Thunor, the Viking deity of lightning and thunder. As a great crowd watched, St. Boniface cut the first notch, and after he had made a superficial wound, the oak's vast bulk is said to have crashed to the ground. Non-Christians were astonished that no thunderbolt from Thunor destroyed St. Boniface, and many willingly converted. Afterward, St. Boniface had a chapel built out of the wood and dedicated it to St. Peter.

Thunor was often worshipped elsewhere in sacred oak groves as well. The fierce red-bearded deity with blazing eyes was brought to England by the Anglo–Saxons and venerated at oak groves in Thundersley in Essex and Thursley in Surrey. The name of Thor, the Norse equivalent of Thunor, has come down to us as *Thursday* and in our word *thunder,* which evokes the sound. Thor rode across the sky in an oak chariot drawn by two goats, and his main weapon was a magic hammer (Mjollnir, the destroyer) made of oak, which sent forth lightning and thunder. Scandinavian temples to Thor housed tall oak pillars, carved to represent the deity.

The renowned social anthropologist James Frazer observes in his master work, *The Golden Bough,* that the reverence paid by the people of Europe to the oak was "derived from the much greater frequency with which the oak appears to be struck by lightning than any other tree of our European forests." Shakespeare wrote of "oak-cleaving thunderbolts," and Russian novelist Leo Tolstoy vividly describes an oak being struck in his classic work *Anna Karenina*: "There was a sudden flash, the whole earth seemed on fire and the vault of heaven cracked overhead."

In 2007, a former Virginia state police official, who had been hit by a lightning bolt, asked me if lightning is attracted to oak trees. He said he had noticed that lightning first struck an oak before striking him and that in hikes across the country he noticed that more oaks are damaged by lightning than any other type of tree.

Intertwined with the myth of the oak and lightning is the story of the mistletoe. Frazer relates the Norse legend of Balder, "the good and beautiful god...the wisest, mildest, best beloved of all the immortals.... Once on a time Balder dreamed heavy dreams which seemed to forbode his death. Thereupon the gods held a council and resolved to make him secure against every danger." Sworn oaths were taken "from fire and water, iron and all metals, stones and earth, from trees, sicknesses and poisons, and from all four-footed beasts, birds, and creeping things, that they would not hurt Balder." Thus, he was considered invulnerable. But one day a mischievous deity named Loki found out that the only creation that had not sworn to spare Balder was the mistletoe, because the sprig was considered "too young to swear."

As long as the mistletoe in which the "flame of lightning smoldered" remained among the boughs, no harm could befall Balder, the good and kindly deity of the oak. But when Loki pulled a piece of mistletoe from an oak tree and tricked a blind deity into throwing it at Balder, the mistletoe killed him, which Frazer explains as a symbolic "death by a stroke of lightning."

Frazer concludes that the reason some peoples, particularly the Druids and Celts who lived in ancient Britain, Ireland, and France, "worshipped a mistletoe-bearing oak above all other trees of the forest was a belief that every such oak had not only been struck by lightning but bore among its branches a visible emanation of the celestial fire; so that in cutting the mistletoe with mystic rites they were securing for themselves all the magical properties of a thunderbolt."

The myth of Balder, Loki, and lightning makes a twentieth-century reappearance in episodes of the *X-Men* comics.

Myths and superstitions connecting lightning to oak trees often assumed a surprising contradictory belief in the Middle Ages, when people kept oak branches in their homes, believing that they would ward off lightning and other natural forces. In some parts of Flanders and France, the remains of the Yule log, frequently made from oak, were regularly kept under the bed as protection against lightning. In England, the Yule log was kept in the fireplace as long as possible, to protect a house against fire and lightning. In some villages of Westphalia, Germany, the Yule log was withdrawn from the fire as soon as it was slightly charred, then replaced on the fire whenever a thunderstorm broke, as people believed that lightning would not strike a house in which the Yule log was smoldering. Eventually, the protective power of the oak was reduced to the symbol of the acorn as a way to keep lightning at bay. A vestige of this tradition continues today with the acorn-shaped knobs at the end of some window-shade and curtain cords.

Other myths, fables, and superstitions about lightning prevailed during the Middle Ages as lightning retained its dangerous and retributive image: Lightning poisons wine; animals killed by lightning should not be eaten; a person killed by lightning should be buried immediately; and lightning will not strike fig trees, grapevines, or onions and will not harm eagles or crocodiles. In the third or fourth century, a young woman named Barbara Dioscorus was martyred by her father, a wealthy heathen, for converting from polytheism to Christianity. Her father was later killed by lightning.

St. Barbara became the patron saint of protection against thunder, lightning, and fire.

*A*mong the many natural phenomena associated with lightning that have ignited superstitions throughout the ages are the displays known to sailors as comazants, corpo santo, or St. Elmo's fire. Described as a kind of ghostly dancing flame of a blue or bluish-white color, or as sort of natural fireworks, the display lasts just a few minutes. St. Elmo, derived from the Italian Sant Ermo or St. Erasmus (ca. 300 B.C.), was the patron saint of early Mediterranean sailors, and the appearance of the flickering light in the heavens was said to signal his benediction to the seamen. St. Elmo's fire before or at the beginning of a storm was interpreted as a sign of the patron saint's protective presence, while its occurrence at the end of a storm was thought to be proof that the sailors' prayers had been answered.

While on his second voyage to the new world, Christopher Columbus himself witnessed the mysterious light, as recorded by his son in *Historia del Almirante*: "During the night of Saturday (October 1493), the thunder and rain being very violent, St. Elmo appeared on the topgallant mast with seven lighted tapers; that is to say, we saw those fires which the sailors believe to proceed from the body of the saint. Immediately all on board began to sing litanies and thanksgivings, for the sailors hold it for certain that, as soon as St. Elmo appears, the danger of the tempest is over." And on that most famously damned of ships, *Moby Dick*'s *Pequod*, Ishmael observes: "All the yard-arms were tipped with a pallid fire; and touched at each tri-pointed lightning-rod-end with three tapering white flames, each of the three tall masts was silently burning in that sulphurous air, like three gigantic wax tapers before an altar."

Scientific explanations of lightning were first suggested in ancient Greece when Aristotle—who believed that thunderbolts were caused by vapors squeezed out from a cloud and that thunder preceded lightning—and others attempted to explain the weather as obedient to natural laws. Even though lightning was widely understood to be a divine phenomenon, a number of early Greek philosophers attempted to explain lightning in scientific ways. Anaximander (ca. 611–547 B.C.) and Anaximenes (ca. 585–528 B.C.) explained that thunder was air pressed against and through clouds, which caused the ignition of lightning. A half century later, Anaxagoras (ca. 499–427 B.C.) thought that thunder resulted from fire flashing through clouds that was then quenched by water in the clouds. In Roman times, the writers Lucretius, Seneca, and Pliny the Elder were among those who continued to try to explain lightning through the workings of natural science. Scientific descriptions about the nature of thunder and lightning continued in the Middle Ages, but as Christianity spread, scientific investigations were swept aside by the teachings of Church fathers and by a rigid adherence to the words of scripture.

The dominant opinion for centuries held that violent weather resulted from either diabolical agency or divine providence, a kind of meteorological struggle between good and evil. Tertullian, an influential ecclesiastical writer in the second and third centuries, believed that lightning was identical to hellfire. This idea was carried from generation to generation of churchmen who found support for Tertullian's view in the sulfurous smell often present during thunderstorms. During the Middle Ages, the diabolical origin of thunderstorms gathered strength as Catholic intellectuals such as St. Thomas Aquinas gave it their sanction. In 1228, when St. Thomas Aquinas was a child, lightning struck the tower where he was sleeping, sparing him but killing a younger sister. From then on, he was frightened of storms and often went to a church when lightning occurred. "Rains and winds, and whatsoever occurs by local impulse alone," he wrote, "can be caused by demons. It is a dogma of faith that the demons can produce wind, storms, and rain of fire from heaven." The chronicles of the Middle Ages are filled with similar notions that storms are caused by a diabolical agency. A strong argument in favor of the diabolical origin was the eccentric characteristics of lightning.

In *A History of the Warfare of Science with Theology in Christendom*, Andrew Dickson White, quoting from a number of medieval sources, observes: "It was said that the lightning strikes the sword in the sheath, gold in the purse, the foot in the shoe, leaving sheath and purse and shoe unharmed; that it consumes a human being internally without injuring the skin; that it destroys nets in the water, but not on the land; that it kills one man, and leaves untouched another standing beside him; that it can tear through a house and enter the earth without moving a stone from its place; that it injures the heart of a tree but not the bark; that wine is poisoned by it, while poisons struck by it lose their venom; that a man's hair may be consumed by it and the man be unhurt."

In a seventeenth-century French account, lightning was said to have left a Latin inscription on its victim and on another "the sacred formula of the consecration of the mass." In a Devonshire church in 1638, noted one report, "extraordinairie lightning came into the church so flaming, that the whole church was presently filled with fire and smoke." Three people died initially and some fifty or sixty were injured, several of whom died later. In one pew two people were badly burned while others were unharmed. And in 1652, during a Sunday service in Cheshire, England, eleven people were killed in a lightning storm. In one part of the church a boy held by his mother was struck dead while his mother was unhurt.

The belief that harmful meteorological phenomena were aimed at the wicked was maintained for centuries. In his influential compilation of moral anecdotes, *Dialogus magnus visionum ac miraculorum*, the thirteenth-century Cistercian monk Caesarius of Heisterbach relates several instances of God using lightning to rescue or punish. He tells, for example, of the steward of his monastery who was saved from a robber by a clap of thunder that frightened the thief and drove him away; how twenty men in a theater were struck down while a priest escaped uninjured, not because he was less of a sinner than the other men but because "the thunderbolt had respect for his profession." Caesarius also tells the story of a priest who was struck in his own church, where he had gone to ring bells against the storm; his sins were revealed by the lightning, which ripped his clothes off and "consumed certain parts of his body showing that the sins for which he was punished were vanity and unchastity."

Medieval religious attitudes toward lightning were summarized by Simon Harward, a Church of England clergyman, who wrote *A Discourse of the Severall Kinds and Causes of Lightnings* in 1607 after a thunderbolt had destroyed a church steeple and bells the previous year. Harward describes three kinds of lightning: judicial, in which "notable offenders are struck down because of their sins"; instructive, which shows God's power and "summons us all to repentance"; and prophetic, which forewarns "of greater calamities to fall afterward upon us unless we amend our wicked and sinful life."

In the early seventeenth century, the Bishop of Voltoraria, in southern Italy, wrote in his *Dies Canicularii* ("Dog Days"), which remained a popular Catholic encyclopedia for over a hundred years, that lightning and thunder were like bombs against the wicked, and said that "of all the instruments of God's vengeance, the thunderbolt is the chief." A few decades later the influential Jesuit Georg Stengel claimed that "the thunderbolt is always the finger of God, which rarely strikes a man save for his sins, and the nature of the special sin thus punished may be inferred from the bodily organs smitten."

While exorcism and prayer were believed to be the best ways to ward off storms directed by God and the devil alike, it was also widely believed that the ringing of bells during a thunderstorm prevented lightning from striking. Some thought that the bells dispersed evil spirits that sought to destroy churches by fire; others claimed that the noise itself disrupted the lightning strikes. In the sixteenth century,

Olaus Magnus, the Archbishop of Upsala and Primate of Sweden, declared it a well-established fact that cities and harvests could be saved from lightning by the ringing of church bells and the burning of consecrated incense accompanied by prayer. He also cautioned that the workings of thunderbolts were to be marveled at rather than inquired into.

That bells could put the hellish legions to flight was widely accepted throughout Europe, even toward the end of the seventeenth century, when the ideas of Sir Isaac Newton would soon turn the laws of nature on their head. In that period, Father Augustin de Angelis, rector of the Clementine College at Rome, published his lectures on meteorology with the approval of the highest Church authority. This learned head of a great college at the heart of Christendom summed up nearly seventeen hundred years of thought when he taught that "the surest remedy against thunder is that which our Holy Mother the Church practices, namely, the ringing of bells when a thunderbolt impends; thence follows a twofold effect, physical and moral—a physical, because the sound variously disturbs and agitates the air, and by agitation disperses the hot exhalations and dispels the thunder; but the moral effect is the more certain, because by the sound the faithful are stirred to pour forth their prayers, by which they win from God the turning away of the thunderbolt."

The beliefs of Father de Angelis and others were reflected in the inscriptions on medieval bells: *Fulgura Frango* ("I break up the lightning flashes"), or *Ego sum qui dissipo tonitrua* ("It is I who dissipate the thunders"). Some bells were engraved with Latin prayers and "baptized" with water from the River Jordan. "Whensoever this bell shall sound," promised one prayer of consecration, "it shall drive away the malign influences of the assailing spirits, the horrors of their apparitions, the rush of whirlwinds, the stroke of lightning, the harm of thunder, the disasters of storms, and all the spirits of the tempest."

Unfortunately, such precautions were likely to invite the very danger they were meant to repel. *A Proof that the Ringing of Bells During Thunderstorms May Be More Dangerous Than Useful*, published in Germany, revealed that from 1750 to 1784, when the book appeared, lightning had struck a total of 386 church towers and killed 103 bell ringers. In their zeal to warn others of the impending danger, the bell ringers themselves often died.

In 1786, the Parliament of Paris issued an edict prohibiting bell ringing during storms. But as late as 1824 four new bells were "baptized against lightning" and placed into the Cathedral of Versailles while government officials took part in the ceremony. In the late 1860s, a visitor in the Tyrolese Alps wrote that the ringing of bells during thunderstorms was an ongoing custom. Even into the twentieth century the bells at a rural church in England might be rung during thunderstorms in the belief that they would overcome the force of lightning.

As lightning figured in the lives of ancient heroes and rulers, it would also play a role in the life stories of several significant figures in Church history, perhaps none more dramatic than in the famous conversion of Paul on the road to Damascus. As the biblical text describes it, the apostle, originally named Saul, was born in Tarsus, in Asia Minor, or modern-day Turkey, "an Israelite of the tribe of Benjamin"

(Philippians 3:5). He studied in Jerusalem under Rabbi Gamaliel, well known in Paul's time, and supported himself during his travels and while preaching, often working as a tentmaker. He was, as he described himself, a persistent persecutor of the early Church, almost all of whose early members were Jewish or Jewish proselytes. He made it his goal to bring Christians to public trial and execution.

One day, as Saul was traveling along the road to Damascus toward Jerusalem, where he was bringing his Christian prisoners for punishment, "suddenly there shined round about him a light from heaven: And he fell to the earth, and heard a voice saying to him, 'Saul, Saul why persecutest thou me?'...And Saul arose from the earth; and when his eyes were opened, he saw no man: but they led him by the hand and brought *him* into Damascus. And he was three days without sight." In Damascus, Ananias, a disciple of Jesus, entered into the house, "and putting his hands on him, said, 'Brother Saul, the Lord, *even* Jesus, that appeared unto thee in the way as thou camest, hath sent me, that thou mightest receive thy sight, and be filled with the Holy Ghost.' And immediately there fell from his eyes as it had been scales: and he received sight forthwith, and arose, and was baptized." After this episode, Saul converted to Christianity, becoming one of the leaders of the early Church and known to us today as St. Paul.

Numerous theories have been proposed to account for his religious conversion, which has been a subject of interest to theologians, philosophers, artists, and physicians for centuries. Some medical experts have suggested that St. Paul may have in fact been struck by a bolt of lightning. His blindness and the return of his vision after three days is consistent with the experiences and injuries of many lightning victims, and his conversion from a persecutor of Christians to a devoted believer reflects the type of change in beliefs, possibly caused by a neurological injury, that some lightning survivors have experienced. Whatever its cause, Paul's conversion was a milestone in the development of the early Christian Church.

Lightning would play another significant role in Church history centuries later in the life of a young German law student named Martin Luther. In 1505, while traveling to Erfurt, in central Germany, after visiting his parents in early July, Luther was surprised by a severe thunderstorm at Stotternheim. Lightning struck nearby, perhaps throwing him to the ground. In fear of death, he called out to Saint Anne, his father's patron saint, with the rather surprising news, "I will become a monk."

Sure that God had sent a sign directly to him, two weeks later Luther honored his solemn promise and entered the Mendicant order of the Augustinian monks. Aware of his escape, Luther proffered the first chapter of the Gospel of St. John as an instrument against thunder and lightning, declaring that he had found the mere sign of the cross with the text "The word was made flesh" sufficient to put storms to flight.

Luther, the founder of Protestantism, challenged the authority of the papacy by emphasizing the Bible as the sole source of religious authority and all baptized Christians as a general priesthood. According to him, salvation was attainable only by a faith unmediated by the Church. His translation of the Bible into the vernacular, making it more accessible to ordinary people, had a tremendous political impact on the Church and on German culture.

*S*ince earliest times, man has tried to explain the awesome flame hurled down from the heavens. Through the structures of magic and superstition, myth and religion, lightning and other natural phenomena were at first attributed to the will, passion, or caprice of great invisible powers and then to God. Observing an inflexible regularity in the order of natural events, as James Frazer concluded, man ultimately sought a solution to the mysteries of the universe in science, and science eventually dispelled most of the myths and superstition regarding nature's forces.

Lightning is still greatly feared, but in recent decades a growing number of daring men and women, curious about extreme weather, have challenged past attitudes and used science as a tool enabling them to find rather than to avoid storms.

STORM CHASER

The east and the west are mine,
and the north and the south
are mine.

—WALT WHITMAN, "SONG OF
THE OPEN ROAD"

or forty-nine weeks a year, David Kearns Hoadley is a mild-mannered retired bureaucrat living out his golden years near the nation's capital, mowing his lawn, shopping at the nearby supermarket, enjoying a quiet life with his wife. During the other three weeks of the year, he chases storms.

Like bungee jumping, skydiving, or mountain climbing, storm chasing is an activity best pursued by adventurous souls who crave jolts of adrenaline. It is a relatively new but increasingly popular pastime in which chasers spend many hours—usually in the spring and early summer, when the atmosphere is most unstable—trying to determine the course of lightning storms and tornadoes and then racing their cars across open spaces to observe or photograph the deadly weather.

"I don't think of myself as doing anything less normal than a hang glider pilot jumping off a two-hundred-foot bluff, depending on a couple of hundred dollars of wire and nylon to save his life," Hoadley says, as if that explains the danger away. "It's not just that the storm is dangerous and exciting in a kind of cheap sense, like going to a stock-car race where the crowd is secretly waiting for disasters," Hoadley once told a reporter. "Part of the satisfaction of chasing is understanding the dynamics of the storm. You're at the right point in time and space out in the middle of the prairie and there's nobody else around. You made the good forecast and everybody else is two states away. The sky gets dark, the prairie grass blows, and you're in fat city, as the chasers say. There's maybe two thousand miles of atmosphere driving this system." He pauses now to measure his words, recalling the moment. "You just stand there and look at it in awe and see yourself as part of a much larger world. It's a kind of poetry in motion that nothing else compares to. You come to appreciate a lot of things, not just the weather."

David Kearns Hoadley is the father of storm chasers. A slightly hunched six-footer with a pale complexion, graying hair, and an obsession with detail, he speaks in the precise tones of a science teacher, peppering his observations with qualifications. He is sixty-nine years old and has been chasing storms for half a century, logging over three quarters of a million miles in pursuit of severe weather. His photographs and writings have appeared in numerous amateur and professional publications, he's been featured in a *National Geographic* special, among other television appearances, and in 2004 his stunning picture of a cloud formation was engraved on a U.S. postage stamp. In 2006, he was honored by the Texas Severe Storms Association, a national organization for training community severe-weather responders, for his fifty years of pursuing storms, and for "his outstanding contribution to severe weather research and education." He was also recognized for starting *Stormtrack*, the chasers' newsletter; for his efforts through it to establish a standard of ethics for storm chasing; and for his gentlemanly character.

A typical chase by Hoadley begins this Wednesday afternoon in the last week of May 2006 with preparations at his home in Falls Church, Virginia. This is peak season in the flatlands from Texas to Iowa, otherwise known as Tornado Alley—a prime breeding ground for many of the eight hundred or more tornadoes and thousands of lightning storms that strike every year from May to July. Hoadley is anxious to be in position for the storms sweeping across the Texas Panhandle. *Position* means being ahead of the storm, usually southeast of the rain-free updraft area. He has invited me to join him in the Midwest.

Ten days before his departure, Hoadley studies computer-model forecasts, looking for an upper-level trough somewhere over the western United States. Troughs are broad, continent-wide waves in the upper atmosphere that bring low-pressure systems, with rising vertical motions, usually leading to clouds and precipitation or to severe weather. As his departure date nears, Hoadley scrutinizes the three-to-seven-day computer models. He packs enough clothes for a week (no time for laundry along the way), two Toshiba laptops (in case one breaks down), national and state road atlases, motel guides, a CB radio (attached to the roof of his car with a magnet), a two-way radio (also attached to the roof), a still camera (Nikon D70), a digital video camera (Sony VX2000), a pencil sharpener, lots of pencils and pens (clipped onto his shirt pocket), an extra pair of eyeglasses (no sunglasses, as they interfere with viewing), a tripod, aluminum foil (to protect recorded video from police radio transmissions), an umbrella ("a professional weatherman always carries an umbrella"), a pillow, cell phone, insect spray, a first-aid and snakebite kit, a wooden board in case he has to change a tire in the mud, an attaché case containing weather maps he has prepared, his checklist of forecasting information, and preprinted maps with the locations of regional weather stations. He will insert his own data on these maps later.

On the day of departure, he waits for the noon version of weather data from the Storm Prediction Center, a division of the National Weather Service. Everything looks good, so he says good-bye to his wife and his friends, who have already wished him bad weather for the trip (he thanked them), and he is off in an Impala by one p.m. Driving through West Virginia, he has doubts about his timing, whether his predictions are accurate. But once he reaches Kansas, his doubts fade into optimistic anticipation. The overnight stop for Wednesday is determined by weather reports. As he drives, he listens to his CB radio to get trucker reports on the "bears" (state police officers who often wear hats similar to Smokey's) and traffic tie-ups to avoid. Light snacks and soft drinks or bottled water take him through the day, and he stops only occasionally for gas or to use a restroom. He avoids heavy meals.

Late that evening, he pulls up at a Super 8 Motel (he is on a tight budget) in Davenport, Iowa. From there he can reach any destination on his storm grid by Thursday midnight or early Friday—Lubbock or Dodge City, North Platte or Aberdeen.

Thursday morning, he is up at seven. No alarm clock. No wake-up calls. He can survive on three hours of sleep. Adrenaline is all he needs.

Before leaving the motel, he draws by hand ("I like the tactile feeling") at least one surface map of weather conditions—depending on how much time he has. Each map takes about an hour to record, plot, and analyze. Storm chasers look for temperature and wind-flow patterns in the atmosphere, which can result in enough moisture, instability, lift, and wind shear to cause tornadoes and severe thunderstorms. To figure out where a thunderstorm will form requires close detective work based on the location, strength, and movement of fronts (transition zones between two different air masses with different temperatures and humidity levels), dry lines (boundaries between moist air masses and drier air masses), and outflow boundaries (masses of cool high-pressure air that descend from an old storm). Some clues are provided by moisture readings and temperatures, near the ground and in the upper atmosphere, that may indicate which storms grow and stay alive. Wind direction and speed at different levels in the atmosphere provide additional evidence of the formation of supercells, severe storms that besides causing tornadoes can produce hail, dangerous wind, severe lightning, and flash floods.

Hoadley checks the latest local and national data on his laptop from the National Weather Service and Accu-Weather, e-mail chatter from other chasers, and at least four computer-graphic models, created mainly by meteorological departments at several universities that predict the weather from twelve hours to ten days ahead. Although the models generally rely on the same data compiled by the National Weather Service from satellites, weather stations, radar, and weather balloons, each has its own particular biases and flaws. Unlike nearly all other chasers, who rely solely on the computer models, Hoadley makes his predictions based on his own laboriously hand-drawn maps. "Many storm chasers just try to apply science, but a lot of it is gut feeling," he says. "That's part of the fun. It's unpredictable. Part of the difficulty of forecasting is deciding how much weight to give to different variables."

Hoadley's skill is completely self-taught. The Weather Service started to make tornado predictions only a few years before Hoadley began chasing storms in 1956, so he had to develop his own method for finding severe storms. Unlike professional meteorologists, he derives his forecasts exclusively from surface data. "I make mental notes but don't incorporate upper-atmosphere forecasts in my analysis," he explains. "Maybe I'm a little old-fashioned in the way I forecast."

A Hoadley weather map is a snapshot of current weather conditions for the geographic area where he intends to hunt, based on a complicated formula worked out over fifty years of chasing. Each map shows the latest hourly surface data from dozens of Weather Service reporting stations. He uses a fourteen-step process based on a number of atmospheric variables. The first step is to look for low-pressure regions driven by upper-air patterns, and temporary conditions that will influence local weather over the following six to nine hours. He then records surface features, including temperatures (the warmer the air the more unstable), dew points (how much water vapor is in the air), wind direction and speed, and air pressure. He connects this data with multicolored isopleths (lines connecting data points of equal value) to reveal the discrete patterns of each (temperature, moisture, and pressure). He analyzes this data, using several interlocking steps, to identify significant features. The key to his analysis is pattern recognition, looking for similarities to weather maps from previous days on which severe weather occurred. His "previous days" include more than thirty years of storm maps.

"Dave does a hand-based analysis," comments Greg Guise, a TV cameraman from Washington, D.C., whom Hoadley introduced to chasing about ten years ago. "He is an artist in a technical field. It's an art very few people have the knowledge to do." Hoadley's predictions about severe weather, mainly tornadoes, are accurate about seventy-five percent of the time. The government's Storm Prediction Center, relied upon by most TV stations, is more accurate. However, once he leaves his data source (a laptop link to the Internet in a local motel), Hoadley does not have access to the center's up-to-the-minute information.

Before Hoadley, the only other storm chaser to achieve any kind of recognition was a Midwestern farmer named Roger Jensen, who began chasing storms in 1953, three years before Hoadley took up the avocation. "When I was going to school up in Fargo, North Dakota," Jensen once told an interviewer, "I had a different interest than anyone else around. Most of the kids were scared of them, but when I was eight, nine, ten years old I just had a fascination about storms." He corresponded with prominent meteorologists about his hobby and traveled around the Great Plains and the national parks, taking photographs of tornadoes and lightning storms, many of which appeared in national magazines. He used a polarizing filter to deepen the blue tones and bring out the crisp cauliflower appearance of a thunderstorm updraft, and his spectacular photograph of a wedge tornado appeared on the cover of the *Bulletin of the American Meteorological Society*, along with his lengthy account of the record lightning storms and floods during the summer of 1975.

Over the next several decades, Jensen took thousands of slides, mostly of lightning, clouds, tornadoes, and sunsets. By 1999, in failing health, he went to live in Texas, where he moved from one nursing home to another until he found a facility with an unobstructed view of the sky, where he could continue to take pictures of raging storms. He died in his sleep in 2001 at the age of sixty-eight. Asked why he photographed storms all his life, he explained, "It's for the awe of what you are seeing. I was born loving storms."

"That man made quite an impression on me," Hoadley later wrote of Jensen. Tim Marshall, a Texas storm chaser, recalled fondly how Jensen, in his final years, left the nursing home (sometimes without permission), carrying a lawn chair and camera, and walked on a prosthetic leg several hundred feet to a nearby open field, where he sat down and waited for storms.

For some forty years, the world of storm chasing was a small and intimate group of scientists, photographers, and those who simply liked the adventure. They were free spirits, drawn to the open spaces. Hoadley says that some chasers were first influenced by the spectacular twister that turns the world around in *The Wizard of Oz*.

By 1997 things had changed. With the use of laptop computers, access to the Internet, and, more recently, Wi-Fi, Doppler radar, satellite receivers, GPS, cell phones, and, most important, the release of the movie *Twister* in 1996, the number of chasers increased dramatically. TV stations put together customized chase units, and currently there are about one hundred fifty to two hundred serious chasers and another estimated two thousand tourists yearly who pay up to $3,000 a week to join a chasing tour.

"Storm chasing has become a whole subculture involving hundreds of people," Hoadley says, removing and wiping his eyeglasses. "Husband-and-wife teams, young and old, scientists and the handicapped, prepaid tours; Americans and foreigners swept along in an expanding hobby with an increasingly rich history. It's one of the last great adventures. I was fortunate to be at the beginning, but it has grown far beyond anything I ever imagined. But something's missing now. It used to be easier to be more romantic about chasing. Today it's more technocratic, more mechanized, more competitive. Maybe at some stage I'll sit in my rocking chair and look at the pictures and video I've taken of great storms and I'll have a great old age."

Based on his maps, Hoadley now thinks that the best chance for severe storms on Thursday will occur in the area around Lafayette, Indiana, about a hundred miles north of Indianapolis and about four hours away from his Super 8 room in Davenport, Iowa. We decide to meet in Lafayette. I had flown out that day from the East. At about three in the afternoon, just as local weather reports on the CB radio from the National Weather Service announce a severe-storm warning with winds of up to ninety miles an hour, a menacing black cloud wall moves from the south. Blinding rain, lightning, and thunder quickly follow. Hoadley is unfazed and keeps driving, tracking the storm. When the rain lets up a bit, he gets out to look at the storm. The wind is strong, making it difficult to open the car doors.

Back behind the wheel, Hoadley turns east, following the storm through Frankfort, Indiana, where a large tree has been knocked down on the sidewalk. Gusts of wind blow through town, swirling papers and kicking up dust. The radio reports that a tractor–trailer in a nearby town has tipped over. But soon the skies clear. The thunder, lightning, and wind are gone.

Once the weather has calmed, we stop for supper at a truck stop and then continue through Indiana and Illinois farming communities as the sun gently sets. Small towns punctuate the monotony of the flat, fecund farms. Watseka, Gilman, Gridley ("home of the Titans") have a recurring sameness—silos, farm-equipment stores, McDonald's, single-level simple houses, trailer camps. A billboard: *"A World at Prayer Is a World at Peace."* Next to it a recruiting poster: *"For honor and country."* Route 80 takes us farther into America's heartland past De Soto, Iowa, the birthplace of John Wayne; past West Branch, Iowa, home of the Herbert Hoover Presidential Library; and finally across the great Mississippi River.

Hoadley can trace his ancestors back to the pre–Revolutionary War era and to a relative who was active in the Underground Railroad in Indiana. His grandfather was an ordained Quaker minister who homesteaded in Kansas in 1914 and died of pneumonia after trying to dig a well by hand during a March thaw. Hoadley's grandmother was one of the first female ordained ministers of any faith. Hoadley's mother was born in 1910 in Stevens County, Kansas, in a sod house.

Hoadley maintains a strong fascination with the sky, land, and history of the West. He reads about and studies the historical significance of places he passes: the Little Big

Horn, the Missouri River route of Lewis and Clark, the Santa Fe Trail. He remembers the "ghosts...the Cheyenne and Sioux; the scout, trader, and pioneer."

Hoadley himself was born in Indianapolis, Indiana, in 1938, and after a few years in Washington, D.C., where his father went to work for the government, the family moved back to Bismarck, North Dakota. There his father took a job as a probate attorney with the Bureau of Indian Affairs, and Hoadley's fascination with storms began. In Bismarck one afternoon at the age of seventeen, he was sitting in a movie theater when he felt a tap on his shoulder and heard his father say, "There's a better show going on outside." A severe thunder and lightning storm had struck, drenching the streets, snapping cottonwood trees and electricity cables. Hoadley and his father drove around town, surveying the damage. "Intersections were a sea of surging water, and great trees had been thrown down on all sides. Most of the city was dark, except for the ghostly blue-white arcing of the power lines, writhing in a tortured dance on wet grass. I was transformed. It was one of those seminal moments, when one turns from a known path and never goes back. My father had brought me out at a critical time in my life and changed me forever.

"Bismarck is my lynchpin," he says proudly. "Everything started there. Sometimes when I'm chasing storms, I'll return to my old high school. I'll sit for fifteen minutes and let the old memories come back."

Soon Hoadley was driving around the North Dakota countryside in the family Oldsmobile, looking for lightning, thunderstorms, and tornadoes. He began storm chasing in 1956, the same year President Eisenhower signed the interstate highway bill, and his friends joke that Hoadley was an "interstate chaser before there were interstates."

Hoadley experimented with his own calculations and observations predicting bad weather, and after a few years they paid off. On a dirt road near Wing, North Dakota, in August 1958, he spotted his first tornado.

"I was a little self-conscious chasing in those early years," he says, "but I wasn't aware I was starting anything new, because there wasn't anything to start! I was just a small-town kid in a prairie state who had a unique hobby. When friends asked my parents where I was on any given summer day, I sometimes wondered how they explained their crazy son."

After graduating from Indiana University in 1960, he briefly attended Yale and then received a master's degree in foreign affairs from the University of Virginia. In 1962 he joined the army. While his buddies eagerly sought assignments in Germany or Japan, Hoadley chose Fort Riley, Kansas. "They didn't realize that was the land of Oz!" he says. While completing his tour of duty in Kansas, he went storm chasing whenever he could. In 1965, thinking he wanted to be an artist or illustrator, he took a course at the Corcoran Gallery in Washington. "Then storm season came along." He quit the course and soon was back on the road (many of his drawings illustrate a book on storm chasing and numerous issues of the *Stormtrack* newsletter).

When he married forty years ago, he warned his wife, Nancy, of his storm-chasing hobby. She accompanied him the first year of their marriage in 1967. "It was a long, hot, dirty ride for her. One day we went through a little town in

Kansas at least three times in three hours. She found that very depressing. The next day she returned home and I remained."

Nancy now takes his hobby in good humor. "No, I don't worry about him," she says. "I always ask before he leaves on a chase, 'Is the insurance paid up?'"

In the early 1970s, he joined the division of water supply and pollution control of the Public Health Service that became part of the Environmental Protection Agency, and he continued to work for the EPA in personnel and budget management until his retirement in 2003. Throughout his bureaucratic career, his desk was often in an office cubicle with no windows. "It felt like a prison," he recalls. "It was confining; I didn't like it." He longed for the open road.

Hoadley decides to spend Thursday night in Coralville, Iowa, in the middle of the state. We pull in to a Super 8 Motel at one a.m. He is up by seven Friday morning and turns on the computer, searching for later severe weather. He tries to finish quickly to get on the road by eight. Every minute counts.

Hoadley hesitates before heading out, when late e-mail messages show that his fellow chasers are congregating in northern North Dakota, where a major storm is predicted. But he believes that it won't arrive until tomorrow. He decides instead to head for southwestern Nebraska to catch a possible tornado (the National Weather Service has predicted a five percent chance, good enough odds for storm chasers to pursue). Driving means nothing to him. On average, he drives about four or five hundred miles each day he is chasing.

More flat farmland, as Iowa merges into Nebraska and Nebraska becomes the Great Plains. The National Weather Service reports tornado warnings to the south on the Kansas border, and Hoadley constantly cranes to stare at the sky from behind the wheel. He sees a string of faint clouds in that direction, but by the time he could get there the storm will have gone. He sees still-fainter clouds on the horizon in the west. Should he stop in Kearney, Nebraska, and take time to study the weather data at a local library where he can plug in his computer? Or should he continue heading toward the Colorado border, where the clouds seem more promising?

He decides to stop in Kearney and check the weather forecasts online. The data models tell him that possible storms are several hours away in the western part of the state, so he continues on to North Platte, Nebraska, where he stops to get gas and observe the sky. It is clear blue with a few wisps of clouds far away in the west.

To Hoadley, a cloud is not simply a cloud. He can spot an approaching storm one hundred miles away when it is merely a hazy, indistinct line on the horizon. When a storm builds up, there may be three or four "cells," or separate systems. He looks for the one with just the right ingredients— a massive thunderstorm, or supercell, which occurs when rising swells of warm moist air push through an overlying stable layer of cooler, drier air. As the air rises, it cools, and the moisture condenses and it forms a cloud. As the warm moist air compresses ahead of the advancing cold front, clouds topped with large cauliflower towers (short for *towering cumulus*, a cloud element with appreciable upward vertical movement) create a chimney effect. As the warm air whirls up, the surrounding cool air sinks. For lightning storms, he looks for cumulus clouds that rise quickly into the upper atmosphere and spread into mushroom-shaped anvils (so called for the flat, spreading top of a cumulonimbus formation—a vertically developed cumulus cloud, often capped by an anvil-shaped cloud; also called a thunderstorm cloud).

Although the exact connection is unclear, Hoadley suggests a relationship between lightning and tornadoes. A tornado, defined as a rapidly spinning column of rising air extending between the base of cumulonimbus clouds and the ground, with winds from fifty to three hundred miles an hour, is almost always accompanied by a severe storm. "The frequency of lightning may be less before a tornado and more after, as the electrical differences build up as the storm disintegrates," Hoadley explains.

Although some chasers track lightning storms and tornadoes equally, to Hoadley and most other chasers, lightning is simply a spectacular and unanticipated by-product of the chase. "When you are driving long distances, you have to give up some things. I am searching for tornadoes but I love looking at lightning, and I'll often pull over to the side of the road to watch it. I remember one day in southwestern Oklahoma when fifty people were parked near me. We had just missed a tornado but we were spellbound. Lightning bolts were going off every two or three seconds. The show lasted for thirty minutes. Pow! Pow! Pow! It was stunning."

He recalls other great lightning storms he has witnessed: south of Bismarck, North Dakota, in July 1962, when continuous lightning sparked for several minutes from the center of one tower at 15,000 feet to 20,000 feet, illuminating the massive anvil and western horizon "as fast as you could snap your fingers, like someone flipping a light switch." Another one was near Sitka, Kansas, in May 1999, when frequent lightning around and inside a tornadic supercell illuminated the entire storm structure—almost like daylight.

In the *Stormtrack* newsletter that Hoadley started, chaser Gene Rhoden described a series of spectacular "Anvil Crawlers," which spread horizontally across the sky like the arteries and veins of some cosmic body: "Picture this: Light to moderate spats of rain crackle against the windshield as you relentlessly press on to find a decent motel. SUDDENLY! As if woven of some strange celestial energy, bright spidery channels of light crawl across a sky of deep azure blue. WHAT A SIGHT!!!" Guise, the TV cameraman and Hoadley's longtime friend, sums up the chaser's attitude: "Lightning is nature's way of talking," he says. "Lightning reflects the many faces of Mother Nature, sometimes even the gentle side."

*N*ow, after three days on the road, Hoadley is getting impatient and a bit frustrated. He uses his cell phone to contact Guise and other friends in North Dakota. Nothing has happened today, but they have good news. "Aaaay," he shouts, putting the phone down. "There's a deep low going into South Dakota. Also lots of juice. Let's get up there. We can make Aberdeen by one a.m. It's only a six-hour drive." We spend a quick half hour in a restaurant to polish off a pepperoni pizza, then return to the road.

More green farmland, then hills with rock and scrub brush. Here and there Black Angus cattle are grazing. We enter South Dakota at the Rosebud Sioux Reservation. On the left side of the road is a new casino offering bingo, black-jack, poker, and a variety of slot machines; on the right side are half a dozen boxcarlike houses with no shrubbery or trees and no apparent attempt to improve their barren grim-ness. Hoadley fiddles with the radio, trying to find 96.1 FM, which every Wednesday and Friday from 6:00 to 6:40 p.m. plays Native American war chants and songs. But it's already past eight.

We stop to take pictures of a magnificent sunset, which inspires Hoadley to discuss a subject he's pondered. Some thirteen years ago he became interested in theology and completed a manuscript entitled "In Search of the Cosmic One." On his long drives, Hoadley, a Unitarian, often won-ders about the subject of his book, the connection between God and man, nature and the afterlife, if any. "I believe that when we die we don't drift into a cosmic void," he argues. "There is no continuity to our consciousness, but there is a continuity to the world that survives us. Experience with storms made me realize that there is something out there, a creative intelligence. But I'm not sure the Something cares about us. I believe there is a Divine Being who is every-where."

He denies any connection, but the same year Hoadley explored these ideas in his manuscript, he had a particu-larly frightening experience with bad weather. He was in southwestern Nebraska with his nineteen-year-old daugh-ter, Sarah, whom he had persuaded to accompany him on one of his storm-chasing jaunts. "I was not a happy camper, because there was no tornado. Then the radio announced a tornado warning. It was after dark. I was hoping to show Sarah a funnel against the background of a lightning flash," he recalls now. "Suddenly I noticed rain sheets approaching like a ghostly shroud and a great sound like rushing water. 'Into the ditch now,' I shouted. We jumped out of the car and left a door open with the dome light on and dove into a ditch. 'Dad, I don't want to die,' Sarah screamed. The developing tornado blew right over our heads. It wiped out two nearby farms. Sarah had nightmares for weeks afterward. She never went chasing with me again."

Lightning poses the greatest danger to storm chasers, Hoadley explains. Lightning and big hailstorms ("which don't give you a real warning and can smash car windshields") are a bigger threat than the tornadoes ("you can see where they are moving"). Several chasers have been struck. Gene Moore, a skilled chaser and weather commentator for an Oklahoma City TV station, has been hit twice, but as far as is known, no storm chaser has ever been killed by lightning. Hoadley and most seasoned chasers (unlike "core punchers," a veteran chaser term for careless local chasers, who often do crazy things) take simple precautions, such as staying in their cars when they are near a severe lightning storm, to reduce the odds of being struck.

Remarkably, even though he has been exposed to hundreds of severe lightning storms and some 182 tornadoes, Hoadley himself has never been injured (though a cousin was struck by lightning several years ago while she was working in a hayfield in Indiana). "When the hair on my arm stands straight up, I either run back to the car—the safest place—or immediately drop to my haunches. If I lie flat on the wet ground it's too dangerous." Hoadley recounts several eerie experiences in which lightning surrounded his car and the radio emitted a zipperlike sound (he utters, "zeezeezeezee"). "It got louder and higher in pitch until the radio was virtually screaming and hissing. This signifies the presence of a strong electrically charged field that is a potential target for a lightning strike."

But Hoadley is afraid of neither lightning nor tornadoes. "I enjoy the experience. What doesn't kill you makes you stronger."

We've left the Rosebud Indian Reservation and are driving along Route 90 in South Dakota, heading east and then north to Aberdeen. The towns flash by only as brief glimpses of artificial lights in the pitch-black night. Hoadley nudges the accelerator to sixty-five, careful not to get caught speeding. A ticket means a costly delay. When one foot tires on the accelerator, he switches to the other foot. At his side is the trusty road atlas.

He turns the CD player on. The stirring strains of Aaron Copland's *Appalachian Spring* and *Rodeo* sweep through the car. "Perfect chase music," he says.

At one-thirty, we finally arrive in Aberdeen, South Dakota. The streets are deserted. We pull in to the familiar surroundings of a Super 8 Motel, the first one ever built.

The next morning Hoadley draws a new weather map based on information from the Internet, e-mails from fellow chasers, reports from current surface observations, the

National Weather Service and AccuWeather. The maps he has drawn on each of the previous three days serve as benchmarks. "All the action today is in North Dakota," he explains. We are on the road by nine-thirty. In Jamestown, a green pickup truck makes a sharp, unexpected turn in front of us. We miss it by a foot. Hoadley is unperturbed, boasting that he's never had an accident (though car accidents are a frequent collateral hazard of chasing).

Hoadley calls his chasing friends again, eager to hear what's brewing ahead. They are waiting in Devils Lake, North Dakota, about fifty miles from the Canadian border. We drive around the lake, where gentle waves lap the shore and families are fishing, a refreshing and unexpected sight after so many miles of dry land.

When we meet up with the rest of the group, a chase car and two chase vans are stationed in the parking lot of a Comfort Inn. One of the vans is manned by three chasers from Holland, professional meteorologists who are spending about three weeks in the United States tracking storms. Their van is equipped with aerials on the roof and a satellite dish, and inside it boasts the latest computer technology, including a TV monitor, shortwave radios, several laptops, and a Global Positioning System. Like many other chasers, they are relying on radar reports and computer models, using Wi-Fi connections to upload and constantly update their predictions.

"We're looking for supercells," one says laconically. "But we like lightning." The skies overhead are clear blue. Clouds are nowhere to be seen. A couple from the University of Nebraska is waiting in the other van. On the back window, scrawled in white chalk, are the words *"Just Married. Tornado or Bust."*

Hoadley greets his friends, Guise and Terry Schenk, a retired fire chief from Orlando, Florida. They immediately ask Hoadley his opinion about the likelihood of bad weather, and he reinforces their computer predictions of a big storm coming. "Isn't it amazing," Hoadley says, "that people come from so many different, faraway places and meet at this one spot."

"I first met Dave in 1995, during a story I was doing," Guise says with a warm smile. "There was a tremendous hailstorm the first day. I followed him on a chase to the Texas Panhandle, and I was hooked." Guise, a thin, wiry man, is wearing shorts and a T-shirt. "I grew up in West Pennsylvania, where hunting was important, tracking animals. Here you are hunting one of nature's elusive animals, an animal that never behaves the same. I love the intensive driving and the interaction with a part of the country most people just fly over. I love coming out to the Midwest. It's soul cleansing."

Full of nervous anticipation, we wait in the parking lot. The sun is strong. We keep waiting. Then Hoadley spies promising clouds in the west. A few minutes later the radar watchers roll down their windows and yell, "Let's go."

"We're going operational," Hoadley shouts, taking a deep breath. "Oh, boy." He swerves back on the road and floors it. The chase is on.

With no current forecast map, Hoadley goes visual: "Good turbulence, breaks over there...not so dark. We've got a full tank of gas; everything is perfect. Now we just have to pick the right storm. There—a tower is going up. We've got to be careful. Everything looks suspicious. There may be another system behind it."

He pulls off the road at an intersection, deciding whether to continue to pursue this promising formation or to wait to see what develops. The chase cars behind us stop. Hoadley confers with them on the two-way radio. He wants to go toward towers and small, early anvils that are forming in the south. His friends want to head north, where another bunch of clouds is forming. Determinedly, Hoadley drives south for twenty miles. No one follows us. He stops, gets out of the car, and studies the sky.

Conditions are changing with amazing rapidity. The gray towers have receded. Suddenly all is sunlight and chirping birds. Hoadley tries to contact the others on the radio. They are out of range. The cell phones don't work either.

Then, in the north, thick clouds form. Hoadley moves everything, except his video and still cameras, from the backseat to the trunk. "Now we're ready for anything."

This is what he has been waiting for. He whips the car around in a U-turn and retraces his steps toward the storm. A gray cloud tower, thousands of feet high, is growing in the north. A black cloud, horizontal and cigar-shaped, hovers between the ground and a larger gray cloud formation above it. Hoadley finally makes radio contact with the others, who have been watching the same storm. Hoadley drives north, hoping to be in a position to take a photo of a tornado with the sun behind it. He stops the car beside open fields and gets out.

The clouds are changing every minute. We are in Cando, North Dakota, about fifty miles from Rugby—the geographic dead center of the North American continent—and about the same distance from the Canadian border. Utter silence. No wind. Even the bugs have vanished. Hoadley watches quietly. This is what he lives for. He has made a good forecast and found the storm. And yet there is also an undercurrent of fear associated with the black cloud now stretched out, snakelike, moving toward us.

The sky grows darker. The winds kick up. The trees are defying the gusting lashes. Lightning flashes brilliantly, followed by cracks of thunder. It suddenly turns cold. The ominous black cloud is only a few miles away. There is a sense of danger in the air.

"Look at that," Hoadley says, pointing to a curling white cone that is beginning to twist down from the black snake. He photographs it with his video camera. Then the white cone vanishes.

To our right, lightning makes electric spiderwebs in the sky. As the seconds pass, the oncoming cloud seems to increase in volume. Then, suddenly, the edge of the black shroud melts away. "That's not a good sign," says Hoadley flatly. "We need a central cone, with vortices. It should be winding toward the fields, sucking huge portions up, spinning them into the darkness, with the sound of an approaching freight train."

But this evening the black snake in the sky slowly dissolves at the edges. "That's it. It's going to pass over us. No more storms tonight."

We drive up the road to where a sheriff's car is parked.

We watch the storm, and then join the others who are parked along an open field. They are frustrated, too, particularly Guise, who has exhausted all his vacation time to chase storms and has come up empty-handed. Jeff Piotrowski, who owns a computer business in Tulsa, Oklahoma, and his wife have joined the group. They have just arrived in their van from northeast North Dakota, where they had wrongly predicted tornadoes. Piotrowski is considered the wild genius of storm chasers, using computer data to the nth degree. He bounds over to the others, who greet him enthusiastically.

Anecdotes are recounted, experiences shared. Everyone is disappointed.

*I*t is seven p.m. Saturday night. Hoadley has promised to drive me the five hundred miles back to Iowa to pick up my car. The others will gather at a restaurant, have a few beers, and swap stories about their triumphs and frustrations. Later in the evening, the photos of other, better storms will appear, and everyone will talk about the ones they caught and the ones that got away.

Leaving his friends, the man who started the whole thing reminisces: 1958, Wing, North Dakota, his first tornado; 1962, Leola, South Dakota, when he briefly filmed for the last five minutes a tornado that had touched ground for an hour ("I nailed that sucker"); 1965, Pratt, Kansas, a tornado so perfectly shaped, it "came down like a painter painted it. Warning sirens from two or three places in the town broke the silence, wailing far off in the distance, out of phase with each other. It was very spooky. I'll remember that as long as I live."

Greg Guise calls on the cell phone. He and the others

were in the restaurant when the sheriff told them to rush outside. A cyclonic rotation that resembled an inverted bowl was circling above. But it didn't qualify as a tornado.

"I'm glad they saw something," Hoadley says matter-of-factly, "but I wouldn't have stayed around for it."

We drive through the vastness of the North Dakota and South Dakota night where odors of rich, newly plowed earth fill the air. We pass De Smet, South Dakota, the childhood home of Laura Ingalls Wilder, author of *Little House on the Prairie*. I am reminded of a woman homesteading in a sod house, of an office worker in a windowless cubicle yearning for open spaces, and of the youthful awakening of the man driving beside me, who pioneered an avocation that hundreds would eventually follow.

Finally, I ask Hoadley why he chases storms. "It is not something that can be answered while waiting for an elevator or in small talk at a cocktail party," he responds slowly. "It touches many levels and requires a measured response."

First he refers to the raw experience of confronting the elemental forces of nature—uncontrolled and unpredictable: "They are awesome, magnificent, dangerous, and picturesque." Next he mentions that the knowledge of storm dynamics and structure sets up a unique relationship between man and nature. "Knowing the turbulent mosaic of wind streams that weave over, around, and through the towering thunderheads and understanding their sources in the great rivers of air that sweep the continent make the observer almost become part of that which is observed—as if by force of will, he could detach himself from the earth and ride the wind into the storm's core. Few experiences in life can compare to standing in the path of a big storm while the thunder grumbles and lightning dazzles the sky. I've seen a lot of Western skies, sunsets, and aurora borealis. I have a deep respect for the beauty of the sky."

He's drawn, too, to the challenge of forecasting storms accurately and consistently: "In a field that is still very evolving, each chaser must draw upon science, experience, and intuition. Every day presents a new puzzle of atmospheric ingredients, different from the day before. There is no textbook for what we do. Even the national Storm Prediction Center misses some big ones."

Then there is the experience of the infinite. "Some of us have a sense of powers that transcend the individual, a sense of the eternal, as when a vertical fifty-thousand-foot wall of clouds glides silently away against a golden setting sun." He will always remember: "Every spring and every chase brings back memories of earlier storms and distant places and the exhilarating sensation you had watching your first twister. You feel as though you have not aged at all but are young and free again like the storms, the lightning, and the winds."

FRANKLIN'S HERETICAL ROD

Eripuit caelo fulmen sceptrumque tyrannis. [He snatched lightning from the sky and the scepter from tyrants.]
—JACQUES TURGOT

*L*ike a river that flows over and around the obstacles in its path, the scientific exploration of lightning rapidly progressed in the eighteenth century despite the impediments imposed by superstition, religion, and the clergy. An unbiased investigation required an empiricist, someone not bound by religious inhibitions, someone like Benjamin Franklin.

The religious hostility toward studying lightning scientifically was mitigated, in part, by the New England theologian Jonathan Edwards, a prominent leader of the "Great Awakening," the religious revival that swept through the American colonies in the late 1730s and early 1740s. The son of a Puritan minister, Jonathan Edwards was born in 1703 in East Windsor, Connecticut. As a child, Edwards studied not only the Bible and Christian theology but also the classics and ancient languages. When he was about ten years old he wrote whimsically about the materiality of the soul and revealed his scientific curiosity at fourteen with an account of the characteristics of the flying spider. At Yale University, he studied theology and philosophy and would go on to write highly influential works in those disciplines, with such weighty titles as *A Faithful Narrative of the Surprising Work of God* (1738), *Some Thoughts Concerning the Present Revival of Religion in New England* (1742), and *A Careful and Strict Inquiry into the Modern Prevailing Notions of that Freedom of the Will, Which is Supposed to be Essential* (1754).

In 1726, Edwards succeeded his grandfather Solomon Stoddard as pastor of the Congregational Church in Northampton, Massachusetts, the largest and most influential church at that time outside of Boston. Preaching the Calvinist doctrine that an individual can discover God's grace without clerical intervention, Edwards emphasized personal experience as a way to spiritual transformation. But his harsh orthodoxy, emphasis on original sin, and the belief that only the elected few would ever reach heaven earned him widespread animosity. Dismissed in 1750 by his congregation in Northampton after he had attempted to impose stricter qualifications for admission to the sacraments, he took up a position as a missionary in Stockbridge, Massachusetts, where he pastored a small congregation of English settlers and more than a hundred Mahican and Mohawk families.

In 1757, upon the death of the Reverend Aaron Burr, who five years before had married Edwards's daughter Esther (and was the father of the future vice-president), Edwards reluctantly agreed to succeed his late son-in-law as the president of the College of New Jersey (now Princeton

University). But almost immediately after being installed as president in February 1758, he was inoculated for smallpox, which was raging in Princeton at the time, and died from the disease on March 22, 1758.

If Edwards is remembered at all today, it is chiefly as a preacher who warned of the perils facing unrepentant sinners. In fact Edwards had a great interest in natural phenomena as well. Scientific discoveries did not threaten his faith, as he rejected any inherent conflict between the spiritual and the material. "All things abroad," he wrote, "the sun, moon and stars, the clouds and sky, the heavens and earth, appear as it were with a cast of divine glory and sweetness upon them."

Edwards wrote about thunder and lightning like a scientist, observing that "the temperature of the air it meets with" is what "renders the path of the lightning so crooked." Anticipating Franklin's studies, he observed that lightning could not be a solid body but is "an almost infinitely fine, combustible matter, that floats in the air, that takes fire by a sudden and mighty fermentation, that is some way promoted by the cool and moisture, and perhaps attraction, of the clouds."

Edwards was a quintessential Puritan, a man whose sermons were filled with fire and brimstone and who was not reluctant to condemn little children to hell. He asked his listeners to renounce all worldly ambitions and to spend as much time as possible in devotion and service to God. Given such ideas, he might have been expected to share the belief of those who saw lightning as a sign to heed God's "terrible remonstrances" to repent. But Edwards thought otherwise; to him, light was a favorite metaphor for God's love.

He was enthralled by the spiritual dimensions of light and interested to know its mechanics. One of his early notebook entries expresses his resolve "to show, from Isaac Newton's principles of Light and Colours, why the sky is blue," and "the sun is not perfectly white." To Edwards, lightning was both an admirable work of nature as well as evidence of God's majesty, and his scientific curiosity led him to view lightning with wonder rather than fear. "And scarce any thing, among all the works of nature, was so delightful to me as thunder and lightning; formerly nothing had been so terrible to me. Before, I used to be uncommonly terrified with thunder...but now, on the contrary, it rejoiced me. I felt God, if I may so speak, at the first appearance of a thunderstorm; and used to take the opportunity, at such times, to fix myself in order to view the clouds, and see the lightnings play."

New scientific discoveries would soon change ideas about natural phenomena, making lightning less capricious and mysterious, subject to observation and explanation. But "rejoicing in the electric shock of thunder," Edwards offered a theological alternative—later shared by other clerics—to the harsh beliefs of the Middle Ages.

*O*n the afternoon of May 10, 1752, a retired French soldier known only as Coiffier stood bravely in a sentry box in the village of Marly, near Paris, while storm clouds threatened overhead. A forty-foot-long iron rod resting on a low wooden platform extended above the sentry box. A glass bottle insulated it, preventing the flow of electrical current, at the base. The wind increased and the thunder rumbled.

Coiffier held a wire that was connected to the ground at one end, so that an electrical charge would dissipate into the earth, and touched the other end to the rod. Sparks flew, and he immediately shouted for someone to fetch the local priest, who, followed by a crowd, ran to him quickly, assuming that he was to administer the last rites. But after seeing that Coiffier was alive, the priest grabbed the wire from the downed soldier with his own hands; again it drew sparks.

The shocked priest and the amazed villagers were the witnesses to the first experiment that proved that lightning was a form of electricity. When Thomas François D'Alibard, the physicist who had staged the experiment, reported the results to the Académie Royale des Sciences in Paris three days later, he credited as his inspiration the work of Benjamin Franklin, whose book *Experiments and Observations on Electricity Made at Philadelphia in America* he had recently translated.

Although Pliny the Elder had written in the first century about the electrostatic properties of amber and a fish called the torpedo that gave off electric shocks, it was not until 1600 when William Gilbert, a physician to Queen Elizabeth, wrote his tome *De Magnete* that there was a better understanding of electrostatic phenomena. Gilbert had found that glass, amber, and other substances attracted feathers and other objects when rubbed. He named the attraction *electric*, from the Greek word for amber, *electron*.

For over a hundred years after Gilbert's investigations, the subject of electricity was generally ignored, until various natural philosophers, including Sir Isaac Newton, paid attention to it. But by the middle of the eighteenth century, electricity was a subject that was receiving a lot of scientific attention.

Philip Dray observes in his biography of Franklin, *Stealing God's Thunder*: "It is not entirely clear what first set Franklin on a course of electrical experimentation, although the appeal of electricity to Franklin (or to anyone fascinated by storms, ocean currents, and movements of warm air in a room) is readily understandable...." Franklin was well aware of the scientific revolution occurring in England. The most notable progress was in the construction of frictional electrical machines, which were cylinders or spheres of glass or sulfur mounted on a hand-rotated axle that rubbed against pads of leather to create static electricity. In 1745 physicist Pieter van Musschenbroek at the University of Leyden in Holland placed iron filings at the bottom of a glass container filled with water whose walls were coated inside and out with tinfoil. One end of a wire reached into the water. When the other end touched the tinfoil on the outer part of the jar, sparks flew. The invention, called the Leyden jar, meant that electricity could be stored and moved. Like their counterparts in England, Franklin and associates staged "electricity parties": They performed tricks with electricity from Leyden jars by making bells and wires jump and further experimented with electricity by slaughtering turkeys with it. But electrical discoveries were still in their infancy. "Electricity is a vast country, of which we know only some bordering provinces," Albrecht von Haller, a Swiss professor at the University of Göttingen in Germany, wrote in a 1745 edition of *The Gentleman's Magazine*, a review of current events.

Drawing on these experiments, Franklin began his own exploration of the relationship between electricity and lightning. In a letter written in April 1749, Franklin first sketched out his theories, suggesting that the water vapors in a cloud could become electrically charged and the positive charges would separate from the negative ones. When such "electrified clouds pass over," he surmised, "high trees, lofty towers, spires, masts of ships . . . draw the electrical fire and the whole cloud discharges." In November of the same year he noted in the journal he kept for his experiments: "The electric fluid is attracted by points. We do not know whether this property is in lightning. But since they agree in all particulars wherein we can already compare them, is it not probable they agree likewise in this? *Let the experiment be made.*"

As news from Europe took a month or two to cross the ocean, Franklin was unaware of D'Alibard's successful trial in France when he set out to conduct his own experiment, now familiar to every schoolchild in America. He had planned to use the newly built spire of Christ Church in Philadelphia as the "point" he needed to entice electricity out of the sky, but construction of the steeple had been delayed. Eager to conduct the experiment during the thunderstorm season, Franklin decided to try a simpler version of the experiment. After it was completed, he wrote to Peter Collinson, an early sponsor in England of Franklin's electrical theories, that a "Machine or Kite" flown in a thunderstorm would draw "Electric Fire from the Clouds to such a degree as to charge a Phial, kindle Spirits, and perform all the other Experiments which are usually done by rubbing a Glass Globe or Tube, and thereby the Sameness of the Electric matter with that of Lightning may be demonstrated."

In June 1752 Franklin and his twenty-one-year-old son, William, set out on a windy afternoon to a location said to be near the current intersection of Ridge Road and

Buttonwood Street in Philadelphia, to test his theories. Though neither Franklin nor his son ever wrote a precise description of the experiment, leading some to doubt whether the episode ever took place, Franklin did relate what happened to Joseph Priestley, a renowned British chemist, whom Franklin first met in London in 1766. Franklin supplied information to Priestley for a book and, according to Franklin biographer Walter Isaacson, vetted Priestley's manuscript.

Franklin's kite had been constructed of a silk handkerchief and two cross sticks of cedar. At the top of the upright stick, he attached a pointed wire that rose about a foot above the wood. A length of twine trailed from the bottom of the kite. A silk ribbon was attached to the twine, with a key dangling at the end. Franklin and his son positioned themselves underneath a shed to avoid getting the ribbon wet as the experiment got under way.

Priestley recorded what happened next in his book *The History and Present State of Electricity*: "The kite being raised, a considerable period of time elapsed before there was any appearance of its being electrified . . . at length, just as he was beginning to despair of his contrivance, he observed some loose threads of the hempen string to stand erect and to avoid one another, just as if they had been suspended on a common conductor. . . . Struck with this promising appearance, he immediately presented his knuckle to the key, and (let the reader judge of the exquisite pleasure he must have felt at that moment) the discovery was complete." According to Priestley's account, Franklin "perceived a very evident electric spark" when he touched the key, and as the experiment continued he collected "electric fire very copiously." Unhurt, though no doubt left somewhat wet from the experience, Franklin had shown that lightning carried a strong electrical charge.

Encouraged by the result of his experiment, Franklin pursued his studies. He "was intrigued," Philip Dray notes, "by the odd unpredictability of lightning" and by the age-old question: Why is one person killed by a bolt while another standing next to him survives? For instance, Franklin's *Pennsylvania Gazette* reported that, in the summer of 1752, "on Sunday . . . a Man was struck dead by the Lightning; that another Man was so stunned that it was some time before he recovr'd; and that a child who sat betwixt the legs of one of them, rece'vd no damage."

Franklin also wondered about more prosaic lightning issues, such as how it snaked its way through objects, concluding that it followed a path of least resistance, seeking out materials that were conductive and avoiding those that were poor conductors of electricity. The reason that damage seemed haphazard was that lightning "will go considerably out of a direct Course for the sake of the Assistance of good Conductors," he wrote in a 1753 letter to Collinson.

His next step was to test his theory by devising a "good conductor" of his own. Lightning is an electrical communication between the sky and the ground. Thus, if one could design an instrument to draw the lightning from a thundercloud and conduct its charge safely to the ground, electrical equilibrium would be restored. Three months after his kite experiment, Franklin attached an iron rod with a sharp pointed tip to the chimney of his house in Philadelphia. The rod was connected to an insulated wire that ran down a stairwell to an iron water pump, where it was grounded.

Inside the house, opposite his bedroom door, the wire was divided; the ends were separated by about six inches. A little bell was on each end, and between the bells a little brass ball was suspended by a silk thread. When clouds passed overhead with electricity in them, the ball played between the bells, striking them.

As Franklin predicted, during a thunderstorm the bells began to tinkle furiously, indicating that his rod had captured electrical current from the lightning and conveyed it safely to the ground. Franklin spread the news of his marvelous invention in his *Poor Richard's Almanac*, sandwiched between notices of upcoming Quaker meetings and court dates: "The Method is this: Provide a small iron Rod (it may be the Rod-iron used by Nailers) but of such a Length, that one end being three or four Feet in the moist Ground, the other may be six or eight feet above the highest Part of the Building. To the upper End of the Rod fasten about a Foot of Brass Wire, the Size of a common Knitting-needle, sharpened to a fine point; the Rod may be secured to the house by a few small Staples. . . . A House thus furnished will not be damaged by Lightning, it being attracted by the points and passing through the metal into the Ground without hurting anybody. Vessels, also, having a sharp-pointed Rod fixed on the Top of their Masts, with a Wire from the Foot of the Rod reaching down, round one of the Shrouds, to the Water, will not be hurt by Lightning."

A number of other notable discoveries about lightning were made during Franklin's time. At the end of the seventeenth century, British physicist Robert Hooke had determined that the duration of the sound of thunder depended on the distance between the lightning strike and the observer, and in 1738, scientist J. N. De L'Isle, serving at the Russian court in St. Petersburg, concluded that thunder could rarely be heard from lightning flashes more than fifteen miles away. In 1770, Italian physicist Giovanni Beccaria built a device called a ceraunograph, which could measure the magnitude of lightning strikes. In Beccaria's instrument, a thin strip of waxed paper was inserted into a gap in a wire connected to a lightning rod. When the rod was struck, the electricity burned a hole in the paper, and the intensity of the discharge could be determined by the size of the hole.

In 1777, Georg Christoph Lichtenberg, a professor of physics at Göttingen University in Germany, developed a more sophisticated way to measure lightning strikes by conducting the electrical discharge of a lightning strike to an insulator covered with a powder. Depending on its intensity, each lightning strike left a subtly different radial impression in the powder. By then pressing blank sheets of paper onto these patterns, Lichtenberg could transfer these fernlike images, which came to be known as Lichtenberg figures, to record the strength of the electrical discharge. Lichtenberg's idea was taken a step further in 1924, when the Westinghouse Company introduced an instrument called the klydonograph, which used photographic film instead of powder to record lightning charges and served as a silent watchman on a vast network of power lines, providing a detailed record of the magnitude and polarity of lightning discharges that struck the lines.

But the greatest triumph of the eighteenth century remained Franklin's so-called lightning rods, which could potentially prevent a great deal of damage to property and loss of lives by blunting lightning's capricious attacks. In the summer of 1752, Franklin's rods were very publicly put into service when they were attached to the Academy of Philadelphia and the Pennsylvania State House, now known as Independence Hall.

*A*s with many advances in science and technology, however, Franklin's lightning rods soon stirred controversy. The prominent French cleric Abbé Jean-Antoine Nollet, a leading electrical scientist and a frequent visitor to Versailles where he was a tutor to Prince Louis-Ferdinand, warned that it was "as impious to ward off Heaven's lightnings as for a child to ward off the chastening rod of its father." Anticipating such criticism from the clergy, Franklin had introduced his providential lightning rods in *Poor Richard's Almanac* in 1753 with the benediction that "It has pleased God in his Goodness to Mankind, at length to discover to them the Means of securing their Habitations and other Buildings from Mischief by Thunder and Lightning." Now, in April 1753, writing to his friend Cadwallader Colden, Franklin quickly offered his rebuttal to Nollet's views: "He speaks as if he thought it Presumption in man to propose guarding himself against the *Thunder of Heaven!* Surely the Thunder of Heaven is no more supernatural than the Rain, Hail or Sunshine of Heaven, against the Inconvenience of which we guard by Roofs & Shades without Scruple."

Of his critics he wrote in a letter to Harvard's John Winthrop in 1768: "It is perhaps not so extraordinary that unlearned men ... should still be prejudiced against the use of [metal] conductors, when we see how long even philosophers, men of extensive science and great ingenuity, can hold against the evidence of new knowledge. They continue to bless the new bells and jangle the old ones whenever it thunders. One would think it was now time to try some other trick; and ours is recommended."

Opening a second front, Nollet challenged Franklin on scientific grounds as well, questioning whether "all these iron points...are more likely to attract lightning than to save us from it." Most damning, however, was his allusion to the unfortunate death of Georg Wilhelm Richmann, who has the distinction of being the first man killed while studying lightning. Richmann, a physicist who was born in Estonia and moved to St. Petersburg, had set up a device called a gnomon or electrometer for measuring the strength of electrical currents in the air. The device consisted of a glass vessel containing brass filings attached to an iron rod that extended from the vessel to the roof of Richmann's house. A thread was tied to the rod, which lay flat until it was charged by the current running along the length of the rod; the degree to which the thread angled away from the rod was a measurement of the intensity of the current.

On August 6, 1753, shortly before noon, pedestrians near Richmann's house observed a dark cloud that "seemed to float very low in the air" and heard "a thunderclap as has hardly been remembered at Petersburg." At that moment, Richmann's assistant in the experiment, an engraver named I. A. Sokolov, saw "a globe of blue fire" jump from the rod of the gnomon toward the head of the professor, who was about a foot away from the rod. Sokolov was knocked down by the lightning and could not remember hearing the clap of thunder, which was "as loud as that of a pistol."

Writing some years later, Joseph Priestley reported

that "an inconsiderable quantity of blood" issued from Richmann's mouth. "There appeared a red spot on the forehead, from which spirited some drops of blood through the pores, without wounding the skin. The shoe belonging to the left foot was burst open, and, uncovering the foot at that place...[was] a blue mark; from which it was concluded, that the electrical force of the thunder, having entered the head, made its way out again at that foot." Abbé Nollet took the opportunity of the unfortunate Richmann's death to argue that attracting lightning by means of a metal rod was a dangerous proposition, to which Franklin responded that Richmann's apparatus had not been properly grounded.

The efficacy of the lightning rod soon spilled over into legal disputes. The summer of 1783, Dray observes, was "a strange and distressing season for contemplating man's relationship to nature. It was a summer of strong atmospheric electricity and a permanent dry fog over much of Europe.... The sun's light was opaque, and diffused by haze...and the air itself seemed fouled." In this atmosphere, a court case was brought against Charles Dominique de Vissery de Bois-Valé of Saint-Omer, France, a retired lawyer and inventor, by his neighbors Mme. Cafieri and her friend Mme. Renard-Debussy, whom Vissery described as "an old quibbler."

Vissery had installed an ingenious lightning rod on his roof consisting of a sword placed on a weather vane. At its base was a globe with an image of lightning. The globe itself was mounted on a sixteen-foot iron bar, the bottom of which was connected to a fifty-seven-foot-long "tail" that went from a wall next to Vissery's house to a rod-and-chain device, which plunged into a well. Mmes. Renard-Debussy and Cafieri's fears of lightning may have been aggravated by "an uncanny occurrence in nearby Arras a few years earlier," when a thunderbolt hit a church steeple with such force that "the flagstones of the church portico were jarred loose and observed to fly through the air." Now Vissery's neighbors insisted he remove the lightning rod from his roof. When Vissery refused, the local magistrate ordered his device to be taken down.

Arguing on behalf of Vissery at the trial was Maximilien Robespierre, a lawyer who would in a few years become a prominent leader of the French Revolution and a participant in its bloody Reign of Terror. In court, Robespierre referred to common sense, praising the utility of lightning rods and disparaging any miraculous powers attributed to what was, in the end, simply a natural phenomenon. "There is no miracle here," he insisted. "That man has dared wrest the lightning from heaven; that he controls all its movement however he chooses ... is also nothing more than a product of the imagination."

Jean-Paul Marat, the scientist, physician, and French Revolutionary writer, was a witness for the prosecution. He argued that the rods were not always capable of drawing lightning from the clouds and recommended as an added precaution that church-bell ringers continue their activities during thunderstorms.

When the victorious Vissery returned home after the trial, he was hardly greeted with a celebration. Instead, when he climbed on his roof to restore his lightning rod, his neighbors cursed him and pelted him with stones. He died the following summer, leaving instructions that whoever should live in his house afterward had to maintain the rod, "as it represented Progress." The new owner, a town official, immediately removed the rod.

In America, warnings against Franklin's invention similar to Nollet's theological objections were gaining ground, especially in the sermons of Thomas Prince, pastor of Boston's Congregationalist Old South Church. Like many others of his time, Prince believed that God made His will known through meteorological events. His view was reinforced in 1746 when his prayers before his congregation, asking God to raise a strong wind to stop a French fleet that was rumored to be on its way to attack Boston, resulted in the wreck of the French ships. Nine years later in 1755, following a series of tremors, Prince published a sermon in which he warned that earthquakes could be caused by lightning. "The more points of Iron erected round the Earth," he wrote, "to draw the electrical Substance out of the Air; the more the Earth must needs be charged with it." Lightning rods offered no protection from the "bitter fruits and tokens of His high resentment of the sins of man. . . . There is no Safety anywhere, but in his almighty Friendship, and by heartily Repenting of every Sin."

"Strange as it may seem to the modern reader," writes historian of science I. Bernard Cohen, "in the eighteenth century the theory that lightning might be related to earthquakes was scientifically acceptable." Prince was challenged by Franklin's friend, Harvard science professor John Winthrop, who in a letter to the *Boston Gazette* echoed Franklin's reply to the Abbé Nollet: "It is as much our duty to secure ourselves against the effects of lightning as against those of rain, snow, and wind, by the means God has put in our hands." Even the young John Adams, the future president of the United States then fresh out of Harvard, reflected on the controversy, noting on his copy of Winthrop's lecture, "This invention of Iron Points to prevent the Danger of Thunder has met with all that Opposition from the Superstitions, Affectations of Piety, and Jealousy of New Inventions, that inoculation to prevent the Danger of the Small Pox, and all other useful discoveries, have met with in all Ages of the World."

While Winthrop appeared to have won the argument on scientific grounds, Prince may have been more successful in influencing public opinion and thereby delaying the installation of lightning rods in Boston.

Franklin took up the matter of the relationship between lightning and divine providence at the Junto, a men's club he had founded in Philadelphia in 1727 to discuss philosophy and other issues of the day. Though the Puritan tradition maintained that God remained closely involved in human affairs, Franklin and other deists believed in a more "general providence," in which God expressed His will through the laws of nature instead of by micromanaging people's lives. Franklin believed in "the existence of the Deity." He held that "the most acceptable service of God was doing good to man." As to God's preoccupation with man: "I imagine it great vanity in me to suppose that the Supremely Perfect does in the least regard such an inconsiderable nothing as man." And at a 1760 Junto meeting, it was mentioned that lightning was surely not simply a manifestation of divine wrath, as "it most frequently wastes itself on inanimate Things as Trees, Houses, etc."

Despite their apparent efficacy, the installation of lightning rods remained controversial, particularly regarding religious structures. Although Franklin's famous invention had been introduced into Italy by physicist Giovanni Beccaria, the famous tower of St. Mark's Basilica in Venice remained unprotected and was badly struck in 1761 and 1762. Not until 1766—fourteen years after Franklin's discovery—was a lightning rod placed upon St. Mark's tower. (It has not been struck since.) In Siena, the beautiful cathedral tower was repeatedly struck until, despite opposition to Franklin's "heretical rod," the tower was at last protected, and though a very heavy bolt passed down the rod in 1777, the church was not damaged in any way.

In Brescia, Italy, however, a lightning strike in August 1769 at the Church of San Nazaro, where large amounts of gunpowder were stored, triggered an explosion that demolished the church, destroyed much of the town, and killed over two thousand people in one of the worst accidents ever caused by lightning. The accident in Brescia helped foment a debate over the use of lightning rods in England, where in 1771 the government asked the Royal Society to recommend the best protection against lightning for ammunition arsenals, including the new royal arsenal at Purfleet on the Thames.

Franklin was appointed by the society to a commission to study the use of rods. The commission fiercely debated whether a pointed rod with its sharp end or a blunt rod with its rounded end was more effective. Though the pointed rods won out, a meetinghouse on the grounds of the Purfleet arsenal outfitted with one of Franklin's pointed rods, some distance away from the gunpowder, was struck and slightly damaged by lightning in 1772. Critics of Franklin blamed the pointed rods.

The dispute on the merits of pointed vs. blunt rods soon spilled over into politics. Because Franklin was also a member of the committee that drafted the Declaration of Independence in July 1776, King George III came to equate the pointed rods with Franklin and the movement for American independence, and he had the offending rods removed from Purfleet, his own palace of St. James, and even the powder magazine stores of the East India Company in far-flung Sumatra. In response, Franklin wrote from Paris, where he was the ambassador for the American Congress now at war with Britain: "I have no private interest in the reception of my inventions by the world, having never made, nor propose to make the least profit by any of them. The King's changing his pointed conductors for blunt ones is therefore a matter of small importance to me. If I had a wish about it, it would be that he had rejected them altogether as ineffectual. For it is only since he thought himself and his family safe from the thunder of Heaven, that he dared to use his own thunder in destroying his innocent subjects."

King George ultimately tried to force the Royal Society to rescind its resolutions supporting the pointed rods, to which Sir John Pringle, the president of the society, replied that it was his duty and inclination to support His Majesty's wishes to the utmost of his power but that he could not reverse the laws and operations of nature. In fact, scientists have since concluded that pointed rods have no appreciable effect on the charge in a cloud, and blunt conductors are every bit as good as pointed ones.

Whether on account of politics, pique, or procrastination, it was not until 1849 that lightning rods were fitted to most British warships. Between 1799 and 1815 alone, lightning damaged 150 ships of the Royal Navy: 18 were set on fire, 70 seamen were killed, and 130 seamen seriously injured. In ten cases the ships were completely disabled and forced to leave their stations during the Napoleonic Wars. Several ships were entirely lost, and at least one, the *Resistance*, blew up from a lightning strike in 1798. The wooden masts themselves were eventually turned into conductors, with overlapping copper strips nailed all the way down each mast to the metal parts of the hull, a technique first adopted by the Russian navy before the British admiralty gave in. With the development of iron vessels from 1860 on, the need to protect wooden ships against lightning declined in importance.

Franklin's rods eventually sprouted on structures throughout America and Europe. By 1870, there were perhaps as many as ten thousand salesmen hawking them in villages across the United States, though perhaps a certain ambivalence remained. In Herman Melville's short story "The Lightning-Rod Man," a salesman tries to persuade a reluctant customer by pointing out that "the oak draws lightning more than any other timber, having iron in solution in its sap. Your floor here seems oak." But Melville's narrator is unconvinced: "I stand at ease in the hands of my God. False negotiator, away!"

With his scientific discoveries, Franklin not only tamed lightning's physical properties but also influenced cultural attitudes. If the divine power of lightning was best epitomized before Franklin in the early-sixteenth-century painting *The Tempest* by Italian Renaissance painter Giorgione, in

Benjamin Franklin drawing electricity from the sky with the help of angels (Benjamin West, ca. 1816).
(The Philadelphia Museum of Art/Art Resource, NY)

which a man and a woman, possibly representing Adam and Eve, anticipate an imminent lightning storm driving them from Eden, after Franklin, lightning may be seen in the service of human progress. In Niquet le Jeune's gravure *Les Droits de l'homme* (1789), a young woman is presenting the Declaration of the Rights of Man to a child. On one side of the image, free citizens dance around a pole topped by a Phrygian cap, which was worn by French revolutionaries, while on the other side a tree struck by lightning falls on a man who represents the feudal aristocracy.

As revolutionary fervor spread, a 1793 French etching, *La Chûte en Masse*, prophesied the fall of despotic rulers, from George III in England to Katherine II of Russia, as a young revolutionary wields "Republican electricity" that overthrows the rulers' thrones. In 1798, in the wake of Napoleon's defeat at the battle of the Nile, James Gillray, a British caricaturist, produced *Destruction of the French Colossus*, which depicts the French Republic as a giant dismembered by British lightning. And as perhaps the ultimate liberating force, in a work by the German painter Johann Heinrich Fussli completed in 1807, the year England abolished its slave trade, a freed slave (Egalite) is entwined with a young girl (Liberte) while they gaze at a slave ship that has been struck by lightning.

An inventor, journalist, printer, diplomat, and statesman, Franklin was the first person living outside England to receive the Copley Medal from the Royal Society of London and was granted honorary degrees from Harvard and Yale. King Louis XV of France wrote a letter praising him. He was one of the original signers of the Declaration of Independence. And yet the iconic image of Franklin is one of a man holding a kite dangling a key from its tail. "At the moment when he drew the electric spark from the cloud," notes historian Andrew Dickson White, "the whole tremendous fabric of theological meteorology reared by the fathers, the popes, the medieval doctors, and the long line of great theologians, Catholic and Protestant, collapsed."

THE ODD COUPLE

Scientists will tell you, push comes to shove, that they really don't know what makes lightning work at all.
— SHERIFF TELLER TO AGENT SCULLY, *THE X-FILES*

artin Uman is an artist. His accomplished oil landscapes and portraits hang on the walls of his spacious office. In blue jeans and tennis sneakers, his gray hair pulled back in a ponytail, he could easily fit the stereotype of the aging bohemian. In addition to his artistic gifts, Uman is also a scientist—the world's leading authority on the phenomenon of lightning.

A distinguished professor in and former chair of the Department of Electrical and Computer Engineering at the University of Florida, and codirector of the International Center for Lightning Research and Testing at Camp Blanding, Florida, Uman has published over one hundred ninety journal articles, written five books on lightning, and received four patents and numerous honors, including the prestigious John Adam Fleming Medal from the American Geophysical Union for "outstanding contribution to the understanding of electricity and magnetism of the earth and its atmosphere."

We are sitting at a conference table in his office at the University of Florida in Gainesville. A portrait he painted of an inquisitive Benjamin Franklin looks down upon us.

"Nothing is really known about lightning," he says. "We know that the light goes here and the light goes there, but why it starts, where it starts—that's very much unknown.

"Understanding really means predicting. You pick a subject, right down to a raindrop. If you look at it in enough detail, you'll find out that people don't really understand it, in the sense that they can't write a mathematical equation to predict what it's going to do. That's understanding."

Uman breaks into an impish grin. In the citation for the Fleming Medal, he is described as "a warm, unselfish individual with a delightful sense of humor."

For the last seventeen years, he has worked closely with Professor Vladimir Rakov, who also teaches at the University of Florida and, along with Uman, directs the Camp Blanding lightning research center. Together, they coauthored *Lightning: Physics and Effects*, a definitive tome. "We argued over almost every word," says Uman. "From my point of view, Vlad is often dogmatic and not willing to listen to reason. From his point of view, I guess, I am some crazy liberal who wants to look at everything seven different ways. In the end he did his part of the book and I did mine." Uman pauses, then says jokingly, "He's very, very smart but too dogmatic. I still don't believe him that the Russians invented radio and TV." Uman and Rakov are the odd couple of lightning research.

Rakov speaks with the slight trace of a Russian accent. With thick wavy hair and a broad forehead, he looks like a

Russian army officer from central casting. He was born in 1955 in Kazakhstan, then a Soviet republic, and completed his formal schooling in the Soviet Union, where he earned his Ph.D. at the Tomsk Polytechnic Institute. In 1986, he was awarded an "Inventor of the USSR" medal and a silver medal a year later from the USSR Exhibition of Technological Achievements. He has lectured around the world.

In the last years of the Cold War, Rakov participated in an exchange program between the Soviet Union and the United States. "We sent nineteen there and they sent nineteen here," says Uman. "Nineteen of our best scientists, including some of our best spies. The Russians did the same. The CIA and the FBI were all over this place. Rakov wasn't a spy, but he couldn't go more than two hundred yards from his apartment." In 1991, Rakov returned to the States on a permanent basis to become a professor in the Department of Electrical and Computer Engineering at the University of Florida.

Uman is seventy-one, twenty-one years older than Rakov. After graduating as valedictorian from his high school in Tampa, Florida, he attended Princeton as an undergraduate and then graduate student, earning a Ph.D. in electrical engineering in 1961. While teaching at the University of Arizona, he became interested in lightning by accident: His first summer in Arizona, he needed a job and went to work for someone at the university who was doing lightning research.

Four years later he joined the Westinghouse Research Laboratories in Pittsburgh, where he expected to do research in plasma physics. But about that time a Boeing 707 was struck by a bolt of lightning while flying over the Eastern Shore of Maryland. It crashed, killing all eighty-one people on board. Soon after, considerable government money became available to study lightning. "Westinghouse formed a group to look at lightning problems, and I was it. Fortuitously, they let me do whatever I wanted as long as I had external—government—financial support." In 1969, he wrote the first technical book on lightning, simply entitled *Lightning*, which tried "to put it all together." In 1972, Uman moved to the University of Florida at Gainesville, enticed by a position to do lightning research. Three years later, he and a former graduate student, Dr. Philip Krider (now a professor of atmospheric sciences at the University of Arizona and also one of the top lightning researchers in the world), after receiving patents on their detection discoveries, collaborated in forming a small company, Lightning Location and Protection, that manufactured lightning-detection equipment, "to make our fortune," he says wryly. "My father, a lawyer, was the secretary. Krider's father, a business executive, was the treasurer. The company was bought and sold a number of times. Krider and I are still consultants." (It did not make either of them rich but did provide some extra income and revolutionized the lightning-research and lightning-protection worlds.) The company is now owned by Vaisala, a Finnish company that makes lightning-detection equipment that's used worldwide.

Despite their differences, Uman and Rakov have made some of the most significant discoveries about lightning of the last fifty years. But to explore their work, it is first necessary to examine briefly the scientific explanations of lightning.

_L_ightning is a powerful electrostatic discharge that occurs in the earth's atmosphere as well as on several other planets, including Venus, Jupiter, and Saturn, and it may have played a part in creating life on earth. The earth's early atmosphere is thought to have consisted of hydrogen, methane, ammonia, and water vapor. When an electric spark of lightning is introduced into this chemical brew, amino acids form, which are the building blocks of protein and primitive forms of life.

Lightning occurs in many forms, the most common being forked lightning, recognizable from both its cartoon depictions and its awesome appearance during a sudden thunderstorm on a hot summer day. Sheet lightning is a shapeless flash of light discharged within and between clouds, sometimes viewed with varying degrees of alarm from the window seat of an airplane.

An estimated 1.2 billion lightning flashes (intracloud plus cloud-to-ground) occur around the world every year. Rwanda has the most; the North and South Poles the fewest. In the United States, there are an estimated 25 million flashes annually.

Florida has the most number of strikes in the United States; Alaska, the fewest. Lightning is responsible for more than ten thousand fires every year in the United States and causes about $5 billion in annual damages.

A typical lightning flash lasts less than half a second, is usually 25,000 feet long, about one to six inches in diameter, and may release about two hundred fifty kilowatt-hours of energy, enough to power a hundred-watt lightbulb for more than three months. It is impractical to try to harness the energy from lightning, however, because the energy in one strike is not sufficient to power a small town for even a day and lightning strikes are too unpredictable.

Most of the energy from a lightning flash is dissipated into the air. This raises the temperature of the lightning channel to as much as 50,000 degrees Fahrenheit, four times hotter than the surface of the sun.

Lightning may strike as far as ten miles ahead of a thunderstorm or while the sky is still blue. No two lightning strikes are the same.

The first process in the generation of lightning is the forcible separation of positive and negative charges, within a mass of air. A substance with a surplus of electrons is said to carry a negative charge, while a substance with a deficit of electrons has a positive charge. (If these terms are confusing, blame Franklin. He coined them.) Electrons naturally flow from an area of oversupply to an area of undersupply, carrying an electrical charge with them.

To see how this results in lightning, imagine a particularly violent seesaw in which a storm cloud is the up side, the earth is the down side. Since the earth is continuously losing electrons into the atmosphere, the bottom of the storm cloud usually contains an abundance of electrons immediately before a lightning flash and is therefore negatively charged. Ultimately, the electrons are released in a powerful discharge in the form of lightning. Like the never-ending motion of the seesaw, the steady loss of electrons into the atmosphere is balanced daily by thousands of lightning strikes that hurl electrons back to the earth.

How this happens is still the subject of research. One widely accepted theory posits that droplets of ice and rain become electrically polarized as they fall through the atmosphere's natural electric field. A related theory suggests that colliding ice particles become charged by electrostatic induction (a method by which an electrically charged entity can create an electrical charge in a second entity, without contact between the two).

The positively charged crystals tend to rise to the top of a cloud and the negatively charged crystals fall to the bottom. Repelled by a group of electrons carrying a strong negative charge in the bottom of the cloud, many of the electrons on the ground move away, meaning that the earth has a deficit of electrons and is therefore positively charged. Tall objects like trees and church steeples are even more positively charged because they are closer to the clouds, the bottoms of which are negatively charged. Tall objects become less negative (more positive) as a result of the repulsive force of the electrons in the clouds.

When sufficient negative and positive charges gather, the electrons probe for an easy path to the ground. Twisting and turning through minor variations in the content and conductivity of the air, the electrons finally reach the earth in the form of an electrical discharge—a lightning bolt. The flood of electrons heats up the air with a violent suddenness, resulting in a sharp increase of temperature and pressure along the lightning channel that causes the air to expand rapidly, like the expansion of gases from an explosion. When this happens, air is pushed aside and set into motion, causing sound waves—thunder.

The sound of thunder might be a sharp crack, if the

lightning strike is nearby, or a rumbling sound due to the varying distance between the listener and different parts of the strike. Another factor that can cause thunder to rumble is the reflection of the sound off clouds, hills, and buildings as the shock waves roll. Since sound travels 1,090 feet per second, or about one mile every five seconds, you can measure how far away a lightning strike is by counting the seconds between the moment you saw the flash and when you heard the thunder, or about five seconds for every mile.

*F*or about a hundred years after Franklin's death in 1790, there was little progress in understanding the properties of lightning. But the last decade of the nineteenth century and the first decades of the twentieth century were among the most fruitful for lightning research, as new photographic and spectroscopic tools were developed to capture lightning in action.

In 1888, M. Hirn, a French engineer, first suggested that thunder is the result of the acoustic shock generated by the extreme heat caused by a lightning flash. In 1895 Aleksandr Popov, a Russian electrical engineer, constructed a device that could register electrical disturbances in the air caused by lightning; and in 1897, Friedrich Pockels, a German physicist, estimated the intensity of lightning current by the use of magnetic fields.

In 1889, with $30,000 provided by the American millionaire industrialist John Jacob Astor, inventor Nikola Tesla, a Serb born in what is today Croatia, built a laboratory in Colorado Springs, Colorado. The lab rose from the prairie floor, with a roof that rolled back to prevent it from catching fire and a wooden tower that rose eighty feet. Inside the wooden structure was an enormous coil, a step-up transformer that magnified power. On the evening of the main experiment, Tesla alerted his mechanic, Czito, to open a switch. A secondary coil began to sparkle and crack, and a blue corona formed around it. Huge arcs of blue electricity snaked up and down the center coil. Bolts of man-made lightning more than a hundred feet long shot out from the mast at the top.

Nikola Tesla's lightning machine in Colorado Springs. This time-exposure photo was made to appear that the inventor was actually sitting beneath the apparatus while the current flashed overhead.
(www.teslasociety.com)

In the same year that Tesla built his monumental tower, a penniless German immigrant named Charles Proteus Steinmetz stepped onto Ellis Island. It was the same year that electric lights were first installed in the White House and that approximately 2,200 lives were lost in the Johnstown Flood. Physically deformed (a diminutive hunchback), Steinmetz was born in Breslau, Prussia, but was forced to flee his homeland due to his revolutionary socialist views. Due to either his appearance or his politics, Steinmetz was at first denied admission to this country, until a friend persuaded immigration authorities to change their minds. The brilliant twenty-four-year-old electrical engineer soon found work in Yonkers, New York, with an electrical manufacturing firm owned by Rudolf Eickemeyer, a German immigrant. General Electric tried to hire Steinmetz away in 1892, but, loyal to Eickemeyer, he refused. Nonplussed, General Electric decided that the only way to secure Steinmetz's services was to buy the entire company.

Steinmetz moved to the General Electric laboratory in Schenectady, New York, in 1893. One of his projects there was to develop "lightning arresters" to halt damage to the high-tension power lines that stretched across the country. When struck by lightning, the lines melted, transformers burned out, and generators were destroyed, and it sometimes took weeks to restore service. Engineers knew that the damage caused by lightning was due to its extremely high electrical voltage, but they didn't know how high. Steinmetz thought that it was not merely a question of high voltage but also of the velocity of the discharge.

The only effective method of testing an arrester was to conduct an experiment with lightning itself. In about 1919 or 1920, Steinmetz built a tower on a remote summit in the

Mohawk Valley in New York to measure strikes, but lightning never struck. However, returning to his nearby cabin in the spring of 1921, he received a pleasant surprise. The front-porch window had been broken. The electrical wires had been severed in a dozen places. His worktable was up-ended and split in half. The icebox was thrown on its side and its legs shorn off. In an opposite wing of the cabin, a large mirror was shattered, its pieces scattered over the floor. "Wonderful...oh, this is wonderful!" Steinmetz exclaimed to an assistant. "That the lightning should have chosen our camp...how lucky we are." A bolt had struck a nearby tree and snaked inside the cabin.

Quite ingeniously, Steinmetz searched out the tiniest slivers of glass and reconstructed the mirror like a jigsaw puzzle, fixing the shards between two large plates of clear glass and sealing the edges with tape. Measuring the pattern of the lightning currents on the fused silvering of the mirror's back, Steinmetz worked out a system to estimate the power of a lightning strike. Eager to continue his research, he convinced General Electric to begin construction of a 100,000-volt generator in Schenectady that would discharge the equivalent of a million-horsepower charge of electrical energy in one hundred-thousandth of a second.

Rebuffing the age-old religious argument that lightning was a divine force and therefore shouldn't be tampered with, Steinmetz said in an address delivered to a Unitarian congregation in Schenectady that "the terror of the thunderstorm led primitive man to the conception of a Supreme Being, whose attribute was the thunderbolt. But when Franklin brought the lightning from the clouds and showed it to be a mere electric spark, when we learned to make lightning harmless by the lightning rod, and finally when we harnessed electricity to do our work, naturally our reverence for the thrower of the thunderbolt decayed." An agnostic, Steinmetz believed, "There can be no scientific foundation of religion. Belief must always remain the foundation of religion, while that of science is logical reasoning from facts, that is, sense perceptions. All that we can say is that the two, science and religion, are not necessarily incompatible but are different and unrelated activities of the human mind."

During the first week of March 1922, dozens of news reporters and honored guests, said to include Thomas Edison, were admitted to Steinmetz's laboratory in Schenectady, where they gazed in wonder at a Frankenstein-like machine. Rising two stories, it looked like a football goalpost with two crossbars instead of one. On these bars were two hundred oblong plates of glass condensers (devices that stored electrical charges). Covered with tinfoil and connected to an electrical current, they were to raise the current to an extremely powerful level. On one side of the machine were rectifiers that converted alternating current to direct current. They resembled oversize electric lightbulbs, while the rest of the contraption was a maze of wires. The source of the energy was the local power system. The voltage of the local system was stepped up through transformers, then rectified through glass vacuum tubes that, in turn, charged the glass plate condensers. Separated by a few feet, two ominous brass domes called sphere gaps were to carry the bolt when it exploded. A block of wood was placed in the gap. The current was to build in the condensers until they all discharged simultaneously with a dramatic flash.

Finally all was ready. Steinmetz nodded his head and an

assistant tripped a switch. There was a small hum as the condensers gathered their charge of electricity.

Suddenly there was a brilliant flash of lightning, followed by a thunderous crash. The block of wood had splintered into tiny fragments. The next day the *New York Times* carried the front-page headline: *"Modern Jove Hurls Lightning at Will."*

Steinmetz had said, "Lightning has always been the most mysterious and most terrifying of all the phenomena of nature." Now he had artificially duplicated a fraction of lightning's voltage and high power and copied the explosive effects of a thunderbolt, enabling him to observe the damage and thereby accomplish the main goal of the experiment—to improve the efficiency of lightning arresters (devices that protected from electrical surges, usually by diverting the electricity to the ground).

But Steinmetz was not satisfied. Estimating that natural lightning was about five hundred times more powerful than his puny creation, he compared the difference between lightning and ordinary electrical current with an analogy. "The pint of gasoline contains more energy and can do more work than a pound of dynamite, but the pint of gasoline gives off its energy slowly, at a moderate rate of power, while the pound of dynamite gives off its energy explosively, all at once, at a tremendous rate of power."

Though Steinmetz was never able to replicate all the force of natural lightning, he used his machine to build and test lightning-proof generators, transformers, and transmission wires. And he ultimately completed the task he had been hired to do, eventually developing an efficient lightning arrester that automatically opened a path by which a lightning bolt could jump harmlessly to the ground instead of entering a power system and melting the equipment. The nationwide power system was no longer at the mercy of lightning.

Stronger generators were later developed, including one that entertained visitors in Steinmetz Hall at the 1939 World's Fair. But it was apparent that laboratory lightning that jumps a few feet cannot be compared to real lightning that leaps many miles. Uman and Rakov are strict constructionists. They believe that the artificial lightning experiments of Tesla and Steinmetz do not add to an understanding of nature's lightning. They do not mention Tesla's or Steinmetz's experiments anywhere in the 688 pages of their book *Lightning: Physics and Effects.*

*T*o most observers, lightning appears to be a single stroke of brilliant light. In fact, we now know that a lightning strike consists of a series of strokes, a significant discovery that was made by Sir Basil Schonland, who is considered by Uman and Rakov to have made the most important contributions to the knowledge of lightning since Franklin. Schonland was born in South Africa in 1896. A brilliant student, he won a number of academic awards and traveled to England to study nuclear physics at Cambridge with some of the leading researchers of the time, including Nobel Prize-winners Ernest Rutherford, John Cockcroft, and C.T.R. Wilson, who, as a lightning researcher, was the first to suggest that thunderstorms feed an electrical charge to the earth. Remote, formal, and austere, Schonland was well suited by temperament to his elite British environment, but he always considered himself first and foremost a South African. Returning to South Africa in the 1920s, Schonland realized that it would be difficult to continue atomic research in such an isolated place, and he decided to take up the study of lightning. His attention was piqued in part by the region's spectacular summer storms—a natural laboratory in the skies that erupted, particularly around Johannesburg, with clockwork regularity on hot summer afternoons.

In one of his early experiments in 1927, he measured the current flowing between the earth and a tree with a galvanometer, an instrument that records electrical activity. He learned that in fair weather electrons flowed from the earth to the tree, but when a thundercloud was overhead, the current changed direction and electrons poured back into the earth. During a thunderstorm, the flow of electricity to the earth was considerably greater than the flow from the earth during fair weather.

In another area of research, Schonland made further advances on the work of C. V. Boys, a British experimental physicist. Boys had invented a camera in which one lens rotated around another stationary lens so that, while the stationary lens took a static picture of an image, the rotating lens captured a series of pictures around the circumference of the same film. Boys had developed the camera in the hope of recording a lightning strike, and he carried his apparatus around for twenty-six years without obtaining a single photograph of lightning. Finally, in 1928, he captured a thunderbolt in New York State with his camera and was able to observe for the first time that a lightning strike consists of several parts: one that progresses downward from the cloud, another that progresses upward from the ground. When he published his findings in *Nature* shortly afterward, he called his discovery "progressive lightning" but attached little importance to it.

By making improvements in Boys's camera, Schonland pursued research for more than a decade, discovering that a single lightning strike actually consists of several elements. First there is a faint, comparatively slow, dartlike downward stroke (called a leader). After touching the ground, it is followed by a more intense and rapidly moving upward main stroke. In other words, a lightning flash, as picturesquely described by author Peter Viemeister in *The Lightning Book*, "starts when a blob of electrons from the cloud gropes toward earth, in a succession of steps . . . creating a faintly luminous trail called the *initial* or *stepped* leader." As it nears the ground, electrons are "sucked" from the air, "creating an ionized *streamer* that rises from the earth to meet the advancing leader. When the two join, the ionized path from cloud to ground is completed and a final leader blazes a faint trail to the earth, charging the air around the path with electrons. Immediately a deluge of electrons pours from this channel of charge, creating the brilliant *main* or *return* stroke that produces most of the light we see." What actually happens is counterintuitive. We think of the main flash as traveling from the sky to the ground, but in fact it travels from the ground to the sky. But the term "main flash" is somewhat misleading. Although the light moves upward, the charge of electrons moves downward. Since everything happens so quickly, the eye sees only one flash. Schonland discovered that when the initial leader is within just 60 to 450 feet of the ground, it "pauses, ever so briefly, like a snake, before striking some elevated object."

Relying on the Boys camera and the cathode-ray oscillo-graph (an electronic instrument that produces instantaneous traces on the screen of a cathode-ray tube corresponding to oscillations of voltage and current), Schonland was able to study a lightning strike in great detail. He measured the velocity of the strokes and discovered that the typical downward leader travels at an average speed of about 250 miles per second, while the main return stroke rebounds at an average speed of about 61,000 miles per second.

In the 1930s, Schonland began to use radio waves to study thunderbolts. (Lightning strikes cause electrons in the atmosphere to oscillate, releasing low-frequency radio waves.) Soon it became clear that what had begun as a method of tracking thunderstorms had a much more practical application. An airplane reflects a radio signal with sufficient intensity not only for its presence to be detected but also for its range and altitude to be measured. This discovery led to secret activity at Orford Ness, along the remote windswept coast of southeast England, where the development of the tracking system that would become radar was unfolding. Schonland was soon asked to participate in the secret research, whereupon he redirected his lightning observatory in Johannesburg toward radar research.

Once World War II began, Schonland returned to England, where he served in the defense ministry, eventually becoming scientific adviser to Field Marshal Bernard Montgomery during the Normandy invasion and its aftermath. Schonland took to the hierarchical structure of the military, and when the war ended, he held a number of high scientific and government posts. In 1958 he became head of Harwell, Britain's leading atomic-energy research establishment. Two years later he was knighted by the queen for his contributions to British science. Apart from his vast research and accomplishments, he developed a number of useful inventions, including an instrument that recorded the increased electrical activity that precedes a lightning strike. This has been used by miners to time the charge of their explosives, helping to prevent many mining disasters.

In retirement, Schonland pursued his lifelong interest in gardening. A habitual smoker, he died at the age of seventy-six in 1972 after suffering a number of cerebral strokes.

While Uman and Rakov agree on Schonland's preeminence in the last two centuries of lightning research, on many other matters they disagree.

Rakov's cramped office is the typical lair of a university professor, overstuffed with books and papers. Uman's office, in contrast, is an expansive lounge, housing a mini-museum of lightning artifacts: a rare poster of a Japanese thunder deity; a twisted radio antenna under which a lifeguard was killed; a blue rocket that was used at the Kennedy Space Center to trigger lightning; a steel plate through which lightning had burned a half-dollar-size hole; many spectacular photos; and several fulgurites, gray and pinkish tubes of silicon glass that resemble tree roots, each about a foot long, formed when lightning burns through sandy soil at exceedingly high temperatures.

With paleontological care, Uman and Rakov recovered the world's longest fulgurite, one branch extending more than sixteen feet, in the sandy soil of northern Florida in

1996. Fulgurites are most commonly found in the United States in Florida, Texas, and the Midwest. "The world is full of fulgurites," says Uman. "All you have to do is go to any beach and start digging. We started digging the fulgurites out of curiosity," Uman remembers. "But they're beautiful. They've become an obsession."

Uman's interest in fulgurites began when he investigated how lightning reaches underground power cables. At the time, it was thought that lightning spread out broadly when it hit the ground. But that's not what happens. In Florida's sand, Uman notes, lightning travels through the ground just as it moves through the air, carving a path that's about an inch in diameter. It melts the sand, and "the sand resolidifies as glass to form fulgurites."

Fulgurites, which are quite fragile, have recently grown in demand among collectors. They are sold on eBay and other Web sites, where a two- to three-inch fulgurite may sell for $100. Many fulgurites for sale—some as much as ten thousand years old—come from the Sahara Desert, where the shifting dunes expose them. Uman sells fulgurites from his Web site to fund his work, and jokes that he'd like to create a diamond fulgurite by putting diamond dust under the right pressure. "Have you ever seen the Neiman Marcus catalog?" he asks wryly. "If we could make a diamond fulgurite, we could put it in the Neiman Marcus catalog. A diamond fulgurite would support our research forever."

Uman and Rakov's investigations are conducted primarily at the International Center for Lightning Research and Testing at Camp Blanding, a hundred-acre facility about thirty miles northeast of Gainesville, which is operated under an agreement with the National Guard. A dirt road across the flat Florida terrain leads past an open gate and a "Military Property No Trespassing" sign. At the end of the road are a mobile home used as an office and a corrugated metal building. Here Uman and Rakov work to artificially trigger bolts from the thunderclouds by means of a rocket-and-wire technique, a high-tech version of Franklin's kite. In *Lightning: Physics and Effects,* Uman and Rakov quote scientist Pierre Hubert: "Triggering lightning at will, at a predetermined place and time, is the old Promethean dream which seems related more to legend than to science."

At the site, slender rockets are launched from steel tubes placed at the top of a five-story tower. When they think the lightning is almost ready to strike, the scientists shoot a rocket about 2,000 feet into the storm clouds. Each rocket trails a thin Kevlar-coated wire to conduct lightning back to a targeted point on the ground. As observers have noted, a leader of lightning suddenly bolts from the cloud, through the rocket, and down the wire, completely vaporizing it with blistering temperatures between 14,000 and 33,000 degrees Fahrenheit. The main lightning bolt—the return stroke—then shoots back into the sky. The display is spectacular. When there is a strong wind, the return stroke moves away from the wire in fiery horizontal strands.

"The big impetus in the States for research into triggering lightning came from trying to understand why Apollo Twelve in 1969 was hit by lightning at five thousand feet and again at thirteen thousand feet," says Uman. "The lightning singed off all of the temperature sensors on the vehicle, and it tripped all of the circuit breakers. They lost all power. But in those days, rockets were hard-programmed to go into orbit. It just went up no matter what you burned up. When they got into orbit, they reset the gyroscopes, the circuit breakers...and they decided that everything was okay, and they went on to land on the moon." Today, commercial airplanes and space vehicles are outfitted with protective equipment. Their shells are generally made of either aluminum (an excellent conductor of electricity) or a composite that contains conductive fibers. If struck, the lightning travels along the exterior shell, then out into the open air.

While lightning research had languished in the decades after the war, money suddenly sprouted with the rise of the space program, particularly after several NASA encounters with lightning. Besides Apollo 12, in 1987 the unmanned Atlas-Centaur 67, which was carrying a naval communications satellite, was struck by a bolt of lightning that altered the memory in the digital flight-control computer and ultimately led to the breakup of the vehicle. Subsequently, the Kennedy Space Center launched a program on triggered lightning.

"But it took several years for people to figure out that Apollo Twelve wasn't struck by lightning—it made the lightning," Uman continues. "When big objects like airplanes and space vehicles get into a cloud, they distort the situation so that lightning is produced even if that cloud was not going to produce lightning by itself. And it turned out that if you triggered lightning and made it come down a wire, you could study its properties. So it was a whole new world. We've been doing it since 1993," explained Uman. "There's a program in Brazil, and Japan and New Mexico as well."

Currently Uman and Rakov's research has two purposes. One is to study the physics of lightning; the other, to study the effect of lightning on things. Among the basic questions researchers are still trying to answer: How is lightning generated in a cloud? How does it strike the ground? Why does it strike what it strikes?

"If you want to stop the lightning, you have to understand the cloud end," Uman says. "If you want to prevent things from getting hit, then you have to understand the ground end. We're looking at everything that happens to lightning close to the ground. We want to understand how it connects, and we're doing a lot with photography and making a lot of electric- and magnetic-field measurements."

Several years ago in the process of an experiment, Uman and Rakov made a particularly important discovery concerning the kind of radiation that lightning emits. "It was really exciting," says Uman. "We didn't expect to see anything at all, and then, all of a sudden, with every lightning stroke, we had X-rays."

Debate over whether lightning emits radiation dates back to the 1920s, when Nobel Prize-winning Scottish physicist C.T.R. Wilson first predicted the phenomenon. He suggested that lightning could produce electrons traveling near the speed of light. Ever since, scientists have presented evidence for and against his theory. Many investigators believed that the lower atmosphere was too dense for electrons to accelerate to speeds high enough to emit X-rays. Instead, they thought that lightning worked by conventional energy discharge—like a spark that occurs when you scuff across a rug. But lately, researchers have measured short bursts of radiation during lightning strikes.

It has long been known that lightning emits radio signals, but the discovery that it also emits waves of energy much higher on the frequency scale—X-rays or possibly gamma rays—was news. "Whether it's an origin of the lightning or a result—that's one of the main questions," says Uman. "The X-ray business is really the hot business now. Scientists are seeing X-rays on satellites that they didn't know were there. It's such a surprise. Apparently they are from runaway electrons."

The leading scientist today pursuing X-ray research is Joseph Dwyer, professor of physics and space sciences at the Florida Institute of Technology in Melbourne, Florida. Instead of chasing natural lightning, Dwyer turned to Uman and Rakov's testing center in rural north Florida, where he's installed a detection system about eighty feet from the tower that measures the flow of X-rays, gamma rays, and other types of energetic radiation emitted by the lightning strokes, all moving at nearly the speed of light. His radiation detectors have recorded energetic radiation in eighty-seven percent of the strokes, occurring at the moment when the charge moved down from the cloud and contacted the ground. "It's right before the visible stroke occurs," says Dwyer. "That appears to be when the energetic radiation is being produced. Nobody really understands completely how this is happening."

In general, the emissions seem to be associated with downward leader steps. They occur in brief microsecond-long bursts, with energies comparable to dental X-rays. Dwyer's next step to determine how lightning produces such high-energy radiation is to make detailed measurements of the X-ray energies and locations of the X-ray sources. To do this, he and his team have constructed a large array of fifty X-ray detectors, covering the whole testing center. Learning more about radiation from lightning could lead to discoveries about how lightning gets started inside thunderstorms, how it propagates miles through the air, and whether or not it is connected to radiation sources from space. Finally, and perhaps most important, discoveries could be forthcoming about radiation's effect on lightning victims.

In his office, Uman explains that his own research is in some ways similar to that being conducted by the experimental physicist Richard Sonnenfeld, a member of the New Mexico Tech faculty and an active researcher at the Langmuir Laboratory. (All of the world's lightning researchers know or know of one another. There are only about two dozen.) In July and August, at the height of the New Mexico monsoon season, researchers there also study lightning by sending objects into the sky.

On a recent late-August afternoon, I drove to the site, about thirty miles southwest of Socorro. Sonnenfeld grew up in New York City, graduated from Princeton and the University of California at Santa Barbara, where he earned his Ph.D. in experimental physics, and went to work for IBM. In a wide-brimmed hat and bushy beard, he has the appearance of a mountain man. "I wanted to do something new with lightning, and I enjoy this kind of outdoor work," he says.

We drive in his white pickup truck past antelope grazing on plains green from fresh rain, past red rocks, across cattle guards, through pine trees, and up a tortuous mountain road with sheer drops of hundreds of feet. Passing from sunshine to fog in less than fifteen minutes, we arrive at the top of South Baldy Peak in the Magdalena Mountains, one of the most likely points to be struck in the Southwest. The Langmuir Lab was put on the top of South Baldy Peak because the afternoon winds sweeping over the nearby plains of San Agustin create storms over the mountain range. We are over 10,000 feet high, and it feels as though we are at the top of the world—nothing is visible above us and the clouds seem close enough to touch. In some ways this is like deep-sea research, probing into the vastness of unknown, unseen regions.

Inside a metal hangar, two helium-filled balloons, each about twenty to thirty feet high, stand waiting like some obedient genii. Attached by fishing wire to each balloon and a parachute below it is a six-pound capsule holding a micro-computer, a sensor, a transmitter, and other instruments. Sonnenfeld describes the capsule as a "little spaceship." The main purpose of the capsule is to measure the electrical field inside a storm cloud, but it also records temperature and pressure readings and other measurements. The balloon and capsule are ready. Sonnenfeld scans the sky, searching for a thundercloud. Only three launches have been made so far this summer. In a separate room of the hangar, Sonnenfeld's students sit beside computers. Suddenly there is a problem. Sonnenfeld rushes to consult with his students at the computers. The computer glitch is fixed. We wait. Still no storms. The end of the day comes without a launch, but Sonnenfeld is philosophical. "Maybe tomorrow," he says.

The few times the balloon and its cargo are released, they can soar as high as ten miles and above into the atmosphere. The balloon slowly expands as it rises. A system cuts the balloon away after a fixed time, or when it has reached an altitude that is above the storm. Then the capsule falls with a parachute to the ground. The catch is retrieving the capsule. To collect the data, the researchers must pinpoint the capsule's location via the Global Positioning System (GPS), wherever it falls—even in hard-to-reach mountain ravines. One capsule landed fifty miles away from the launch site in Truth or Consequences, New Mexico.

"My dream," says Sonnenfeld, "is to be able to tell you where every charge in a lightning strike came from and where it went."

New Mexico Tech and its Langmuir Laboratory developed the Lightning Mapping Array, a three-dimensional location system that measures the radiation from a lightning discharge at multiple stations and locates the sources of the radiation to produce a map of lightning activity. In addition to the Langmuir Lab in New Mexico, Lightning Mapping Arrays (LMAs) are currently in operation in northern Alabama, the White Sands Missile Range in New Mexico, in Washington, D.C., and in Oklahoma. The Oklahoma LMA consists of a central analyzer in the city of Norman and ten stations distributed across the central part of the state. Each station measures the time at which a radio pulse arrives from the lightning that generated it. From the differences in times that the pulse arrives at seven or more stations, the system determines the time and location at which the lightning formed. Up to thousands of segments can be located for each lightning flash to reveal its initiation and development inside storms.

Cloud-to-ground lightning data has been collected in real time since the late 1970s. The first uses were for forest-fire detection and utilities. Other groups have found the network data useful in aerospace and military operations, explosives handling, aviation operations, communications, and meteorological research and applications. The principal system for tracking lightning is the National Lightning Detection Network, owned and operated by the Vaisala company, which bought Uman and Krider's company, and which has over one hundred ground-based sensing stations located across the United States that instantaneously detect the electromagnetic signals given off when lightning strikes the earth's surface. These remote sensors send the raw data via a satellite-based communications network to receiving stations.

At the ultramodern National Weather Center building (completed in 2006 and funded in large part by the federal government) at the University of Oklahoma in Norman, Oklahoma, a receiving station displays on a fifty-inch plasma screen tiny lights and then records on computers about ninety percent of lightning flashes across the United States almost immediately as they occur. By at least the year 2020, experts at the center predict that satellites will similarly measure and record lightning strikes from around the world. Researchers Donald MacGorman and David Rust at the center and its National Severe Storms Laboratory are also conducting research to determine where lightning develops and where it goes in a cloud. MacGorman, who "started out in space physics but was always fascinated by lightning," said that a basic question about lightning is still unknown: "How does it start?"

The mysterious St. Elmo's fire that seemed so portentous to Mediterranean sailors is now understood to be caused by a discharge of electrons from the strong electromagnetic field induced during a lightning storm. The glow is actually ionized air that frequently appears to settle on the masthead of vessels in warm weather, and especially in hot climates, and is considered an electrical phenomenon, though it is never known to produce any of the disastrous effects of lightning. When it is confined to the topmast, it is considered a prognostic of bad weather, though not to such an extent as to cause injury. But when it descends down the mast, it is believed to be a sure proof that a storm is imminent. With the age of flight, St. Elmo's fire has appeared along the wingtips, propellers, and antennae of aircraft, often disrupting radio communications. There is even a theory that the *Hindenburg* airship fire and subsequent crash in 1937 in New Jersey, which killed thirty-six people, may have been sparked by St. Elmo's fire igniting hydrogen leaking from the dirigible.

Another strange phenomenon, which has been observed around the world since Aristotle first described it in the fourth century B.C., is the mysterious phenomenon known as ball lightning. Nearly everything that is known about it today has been deduced from reported observations. While very few who see it ever forget it, no video or motion picture of ball lightning has ever been taken.

As its name suggests, ball lightning is a luminous, spherical glow ranging in size from a golf ball to a basketball that most often looks red but can also be blue, white, or green. It usually appears during heavy thunderstorm activity or soon after a nearby cloud-to-ground lightning flash. Sometimes, however, it appears near or in clouds without the apparent presence of lightning, and sometimes such objects are reported in seemingly clear air.

A family's reaction as ball lightning passes through their home
(from *The Aerial World* by G. Hartwig, 1886).
(Courtesy of The National Oceanic and Atmospheric Administration)

Ball lightning appears to move horizontally at a speed of at least a few yards a second, but it may also seem to stop and change direction. While ordinary lightning lasts about a tenth of a second, ball lightning may flash for several seconds and sometimes even minutes. It may be attached to a conductor or be free-floating, and the balls that hover near the ground may act differently than those that hang high in the air. It has been said to enter rooms through windows and chimneys. Many observers report an accompanying odor, which is usually described as sharp and repugnant, resembling burning sulfur or nitric oxide.

The characteristics of ball lightning are not understood, and its very existence has often been debated. In 1936, W. J. Humphreys, a physicist and meteorologist for the U.S. Weather Bureau, after collecting data on 280 reported sightings, concluded that not a single case truly described a "ball" of lightning. In *The Lightning Book*, Peter Viemeister writes about the experience of his friend E. Markow, a graduate engineer, in the summer of 1943. "He was upstairs taking a shower in his frame house," he writes. "A brief downpour had just passed, and he had opened the windows at both ends of the hall. He had finished his shower, heard a peal of thunder and was about to step into the hall when he saw a bluish ball, about 12 to 18 inches in diameter, float through the screened window at the end of the hall and start to come toward him. Startled, he stepped back, and the ball, seeming to move at about the speed of a gentle breeze, drifted by, about waist high...the glowing ball...proceeded down the 30 foot hall in about 3 or 4 seconds, and passed out through the screened window at the other end, where it dropped out of sight without any noise."

In another remarkable account, a ball-like flash of lightning entered the home of Jean Zaleski in New Hyde Park on Long Island over fifty years ago. A tiny, red-haired painter, now in her eighties, Zaleski has displayed her canvases at the Metropolitan Museum of Art, the National Academy of Art in New York, and the Palazzo Vecchio in Florence. To this day, Zaleski's experience sends shivers down her spine. She and her three children were swimming in the pool outside their house. Seeing a storm coming, she rushed the children inside, where they all went upstairs to change out of their bathing suits. All the windows in the house were open.

"Suddenly, a flash entered the room, and I yelled at the children to get on the bed. All four of us jumped on it," she told me in a recent interview. "The children's bathing suits were wet and left puddles on the floor. The lightning hopped from one puddle to another and then it went out again. It never hurt us, but I've never forgotten it." The flash, said Zaleski, was "the scariest thing that ever happened in my life."

Martin Uman himself claims that he has "personally received over one hundred unsolicited eyewitness accounts of ball lightning," and he has reported a number of credible stories. "Unfortunately," he writes in his book *All About Lightning*, "there is at present no adequate theory of ball lightning. For example, no theory can account simultaneously for the degree of mobility, the constancy of light output, and for the fact that the ball does not rise. Despite numerous theoretical models proposed, the causal mechanisms remain unknown. All theories fall into one of two general classes: those in which the energy source for the ball is postulated to be outside the ball (externally powered ball lightning) and those in which the sustaining energy is postulated to be stored within the ball itself (internally powered ball lightning)."

Seeking to explain two mysterious phenomena in one fell swoop, Uman suggests that "a small percentage of the UFO reports are so similar to a certain class of ball lightning that they both must refer to the 'same imperfectly understood physical phenomena.' These particular lightning balls seen in the sky are much larger, much brighter, and much longer-lived than the typical balls. They are reported to be about 10 to 20 feet in diameter, give the impression of being as bright as lightning, and may last for a minute or more. When such objects appear immediately after lightning, it is clear that they should be called ball lightning."

More recently, strange bright objects spotted high above the clouds—low-luminosity lightning discharges—have come under increased scientific observation. Flashes of light that fill the night sky, they are largely unpredictable, short-lived, and inherently difficult to study. In the U.S., they occur most frequently over the high plains, usually within a sixty-mile radius of a lightning strike. They have been reported for over a century, but prior to the scientific recordings that began in the 1990s, they were generally dismissed. Now scientists have given these creatures the fanciful names of sprites, blue jets, and elves.

Reddish-orange blobs resembling jellyfish, which appeared for a mere few milliseconds over active thunderclouds some forty-five miles above the earth, were first photographed in 1989. Dr. Davis Sentman of the University of Alaska, one of the few scientists studying these luminous, ghostlike phenomena, named the eerie flashes of colored lights after Shakespeare's mischievous spirits of the air— Ariel in *The Tempest* and Puck in *A Midsummer Night's Dream*. The cones in the retina of the human eye, which see color, can barely perceive the light of these sprites, but the more-sensitive achromatic rods of the eye that permit night vision can register their low-light level more easily. Low-light cameras on airplanes and the space shuttle have recently captured more images on film.

No one is sure what sprites really are or what causes them, although researchers believe that they may contain a great deal of energy. "Although we're not certain," said

Sentman, "we suspect that the energies from sprites may be sufficient to drive some novel chemical reactions. The region of the atmosphere where sprites appear typically doesn't contain a lot of energy, so the energy produced from a sprite could do some really interesting things." They have strong electric fields and electromagnetic pulses that may interact with the earth's ionosphere (the uppermost part of the atmosphere, which is ionized—gaining or losing electrical charges—by radiation from the sun) and magnetosphere (a region surrounding the earth in which the motion of charged particles is dominated by the earth's magnetic field), and researchers have speculated that they may play a role in the formation of the earth's ozone layer.

Blue jets, which are brighter than sprites, project from the top of a cumulonimbus cloud, typically in a narrow cone to the lowest levels of the ionosphere, twenty-five to thirty miles above the earth. Purplish blue in color, they were first recorded in 1989 on a video taken from the space shuttle as it passed over Australia. In 2002, five gigantic jets between thirty-five and forty-five miles in length were observed over the South China Sea. The jets lasted under a second and were shaped like giant trees and carrots.

During the space shuttle mission in 1990, another type of light—a dim, flattened, expanding glow that hovered about sixty miles above a thunderstorm—was recorded off French Guiana. As more of these lights were observed, they were nicknamed elves—for Emissions of Light and Very Low Frequency Perturbations from Electromagnetic Pulse Sources—which refers to the process by which their light is believed to be generated. Their color was a puzzle for a time, but it is now believed to be a red hue.

Finally, when asked to explain the phenomenon of heat lightning, Uman and Rakov dismiss it as nothing more than the faint flashes of lightning on the horizon or in clouds from distant thunderstorms that are often too far away for the thunder to be heard. Heat lightning is so named because it often occurs on hot summer nights, and it can be an early warning sign that thunderstorms are approaching.

Wrapping up our conversation, Uman glances at his watch. "I have an appointment with the dentist," he says.

But I have one more question. "Have you ever been struck?"

"Never. My wife is scared to death of lightning. She won't even let me go outside to watch it."

THE MOUNTAIN CLIMBER:

THE SUMMIT

*The spirit didn't come so far to
slip all down to nothing....*
—GEORGE LEIGH MALLORY,
FROM R. L. G. IRVING'S
"SOLVITUR IN EXCELSIS"

*T*he thirteen climbers who made up Rod Liberal's party
in July 2003 were mainly in their twenties and thirties and
were friends, family, and coworkers in the technology department at the Melaleuca company, an alternative-healthcare firm based in Idaho Falls. All avid athletes, they skied,
played hockey, mountain biked, and climbed. Their life
was the outdoors. For this ascent, they had split up into
four teams, climbing veterans and novices in each group,
and would space themselves across several hundred feet of
mountain.

The first night Rod and his friends had camped at the
Meadows near the base of the Middle Teton and talked until
about three or four in the morning. Following an unspoken
rule in the Tetons, most hikers break camp at around two
a.m. in the summer to reach the summit by noon, to avoid
late-afternoon thunderstorms. But the campers hadn't
gotten up until about seven a.m., and a party of thirteen

climbers, some with little experience, does not move quickly. Such a late departure would surely put at least some members of the group near the summit very late in the afternoon, even if other climbers in front of them did not delay their progress.

The morning of July 26 was beautiful but chilly. Rod was wearing pants that could be converted into shorts, a fleece sweater, climbing shoes, a helmet, and a winter hat with flaps. All the climbers carried radios. Some members of the group had camped higher the night before, and Rod's group met them that morning by the fixed rope, an approximately forty-foot-long rope that rangers had attached to a wall to help climbers. "At that point everything just gets exceptionally beautiful," Rod remembers. "It's amazing being up there. You can see east as far as your eyes can reach and west as far as your eyes can reach. The mood is great. We're all just happy to be there. Nobody was showing signs of fatigue. I'd hike with some friends for a while, then I'd wait up for [the others] to chat with them for a while. It kinda goes that way and then it starts getting pretty technical. There are some areas you have to scramble up, and it's mostly shade, so it's pretty chilly."

*T*he Grand Teton National Park, at the northern edge of the Rocky Mountains, covers about 485 square miles in northwestern Wyoming, near the Idaho border, just south of Yellowstone National Park. When the French trappers of the Hudson's Bay Company came upon the mountains in the early nineteenth century, they called them *Les Trois Tetons*, or "the Three Breasts," a name that stuck and by which the South, Middle, and Grand Teton are still known. The highest peak, the Grand, 13,770 feet high, is bounded on the south by the lower saddle (11,600 feet), which separates it from the Middle Teton. Winds howl across this broad stretch from which most of the summit climbs to the Grand are launched. The higher Gunsight Notch (12,600 feet) on the north separates the Grand from its neighbor Mount Owen. The abrupt vertical rise of the Tetons contrasts with the horizontal sage-covered valley and glacial lakes at their base, creating a spectacular panorama—like a scene from an iconic photograph or a Technicolor Western.

The jagged granite peaks of the Teton Range rise about 7,000 feet above the valley called Jackson Hole, pushed skyward by powerful forces in the earth that erupted eons ago. The Teton peaks are crafted from a block of metamorphic and igneous rocks that formed the central core of the continent, probably deposited about 2.7 billion years ago along a chain of volcanic islands. These rocks were once covered by much younger sedimentary rocks that have since been worn away by erosion. The Tetons record about seven-eighths of all geologic time and hold clues to the history of the birth of the North American continent.

Nathaniel Langford, an Internal Revenue collector, and James Stevenson, of the U.S. Geological Survey, were allegedly the first to scale the Grand, on July 22, 1872, but as is typical of many famous climbs, there is controversy over whether they actually reached the summit. The first verifiable ascent to the summit was made in 1898 by William Owen and Franklin Spalding, members of the Rocky Mountain Club (formerly the Rocky Mountain Climbers Club, established in 1896 in Denver, Colorado), and a prominent route on the mountain is named after them. Twenty-five years passed before anyone climbed to the summit again; three students from Montana State College made the ascent and descent in a single day in 1923, without the use of a rope, under the leadership of geologist Quin Blackburn, who later served with Admiral Byrd in the Antarctic. Since then, the Grand has become one of the most popular climbs in the country, with some ninety routes and variations to the summit.

Most climbs begin in Garnet Canyon, the principal approach to the South, Middle, and Grand Tetons. Access to Garnet Canyon begins at the Lupine Meadows trailhead, which is reached by driving about a quarter mile south of the Jenny Lake campground. From Garnet Canyon, most climbers ascend to the Lower Saddle, considered the best camping area for climbs to the summit of the Grand, about 2,200 feet above. Here there is water from glacial-melt at the left of the trail just near the top of the saddle; it is the last water on the climb. From the Lower Saddle, the Exum Ridge route is the most popular, with its southern exposure, moderate climbing conditions, and familiar landmarks. The trail leads past the Black Dike, beyond which lies a large, smooth tower called the Needle; then a tunnel called the Eye of the Needle leads under a huge boulder. After passing through various gulleys and chimneys (vertical rock clefts large enough for a climber to squeeze through), the route opens out onto a narrow ledge known as Wall Street, named by a guide in the 1930s after a client who was a Wall Street banker.

*R*od's group reached Wall Street, on the southeastern face of the Grand, by around eleven a.m. The ledge extends like a catwalk along the mountain's near-vertical flanks and appears to intersect with a massive ridge that ladders up toward the summit. But about five feet before the ridge, the Wall Street ledge fractures into a 1,000-foot drop. Wall Street is the mountain's equivalent of turning onto a one-way street. Prior to crossing the gap it's possible to turn around and go back down, but once you cross the void it's better to continue to the summit and descend by another route.

In 1931 a young man named Glenn Exum, a gifted climber and superb athlete, first explored the Grand's south ridge, now named in his honor. He was alone and unroped, wearing a borrowed pair of leather football cleats. Exum could see no way to cross the wide gap at the end of Wall Street, so he jumped across from a standing start. Committed then to the ridge (it was impossible to climb down what he had just jumped), he continued up to the summit. That day in the summit register, Exum wrote that the route traversed "the thousand thicknesses of hell."

At Wall Street, Rod snapped a photo of his group while they settled in for the two-hour wait for the climbers ahead to clear. Clinton and Erica Summers and Rod's good friend Jake Bancroft were in one group, but since it was Jake's first time, Erica felt a little uneasy. She asked the more-experienced Rod to switch teams with Jake, so Rod—in an apparent twist of fate—joined Clinton and Erica, and Jake dropped back into the fourth team behind them.

Rod had met Erica for the first time that morning. Erica Esplin was born in Blackfoot, Idaho, in 1977, and had earned her legal secretary's degree from Idaho State University. In 1998, she had married Clinton Summers in the Idaho Falls temple of the Church of Jesus Christ of Latter-day Saints. Like Rod, Clinton was a computer programmer. Quiet, soft-spoken, with a smile always playing about his face, he was about Rod's height of five ten but a bit stockier, with short, receding hair. Erica, with long brown hair and a wide-open smile, was active in the Mormon Church and a devoted mother to her two young children. At the time of the climb, she was taking classes at night to become a registered nurse.

Although Erica was an active water-skier and snow-boarder, she became tired after the wait at Wall Street and didn't know if she could go on. Rod and Clinton "sassed her and tried to motivate her," Rod recalls. "She was being a good sport about it. She told us, 'My feet hurt, but let's go.' But she wasn't cracking jokes anymore. She was all business. I think most of us were."

Rod's group made it past Wall Street and then encountered "one of the first really scary slanting ledges, where you face a thousand-foot drop to your right. There's a big boulder in the way, and the only way to get past it is to climb around it. You're roped in, but behind you there's nothing—only a free fall. You have to kinda hold on and do this short leap of faith.

"As we got closer to the summit, a little Cessna flew by right below us. We were waving and they were waving. As you got higher you could see Idaho and Wyoming. It was amazing being up there. Nothing around you is even close." They had a hawk's-eye view of the snow-tipped summits of the Tetons.

The three finally got to the base of Friction Pitch, where a boulder-covered ledge leads steeply up to a blackrock gully and a small notch at the ledge above. While they were contemplating their ascent, the climbers got a call on the radio from Robert Thomas, who was in the group ahead. "You guys can't see this," he related, "but the weather is looking pretty grim." It had just sprinkled a little bit and everyone was wearing raincoats and hoods. Robert said that if everyone agreed, the group was going to ditch the summit and just try to reach the next belay area and start heading back down the mountain. The three saw a storm moving in to the east and "everybody's mood just changed."

Friction Pitch was arguably the most difficult part of the Exum route to the summit, but the three had no choice but to continue. The safest way forward was to climb one at a time—one person leads the climb to the top, gets anchored in, and then belays the next person up. But to save time, instead of each climber pulling the next one, Clinton brought the rope up for the remainder of the group.

As Rod and the Summers couple prepared to scale the wall, two other teams were already ahead of them above Friction Pitch. With the rain blowing in, the leaders of team one had decided to skip the summit and follow a different route to the west. They had already begun to rappel down the cliffs. The second team, including Robert Thomas and his father, Bob Thomas, was spread out above the top of Friction Pitch and was headed toward the rappel point. The fourth team was behind them, waiting for Rod's group to move.

"So Clinton went up and got anchored in," Rod says. "Then Erica went up. She got anchored in and sat down next to him. Then it was my turn to go up. I was wearing a climbing harness the whole time. It goes around your waist and two straps go around your legs. The belay device attaches to a loop on the harness that creates friction in the rope as it runs through it. Clinton had the rope through his device and he had another system of ropes attached to three anchor points, devices that squeeze into cracks and that have large teeth that hold inside the cracks. So he is anchored in and I am basically anchored to him. Erica was also anchored to him, sitting to his left. Both were sitting on solid rock."

When the lightning lances the mountain, the first team of climbers is high enough above to avoid injury. The second team is directly above Rod's group. The team includes Bob Thomas, who made the first call to the rangers, and his son Robert, who loses consciousness, spins around, and tumbles toward his wife, who stops his fall. As he comes to, dazed, he pulls the aluminum trekking poles off his back, fearing they will attract another strike. Then he hears a bone-chilling scream and recognizes the voice of his friend Clinton Summers.

Robert creeps past an outcropping and along a narrow ledge toward where Clinton is sitting with his wife, Erica, leaning heavily against him. She doesn't move as Clinton's screams continue to echo. Robert crawls over the ledge and slids in next to Erica, whose body falls across his lap. "The right side of her face and body were burned horrifically," Thomas later recalls.

Besides the Summerses, the other member of the third team is Rod Liberal. Hit directly, he is dangling in a contorted position some 13,000 feet in the air, held only by one rope.

The fourth team of three climbers, below, includes Jake

Bancroft, who had earlier switched groups with Rod. They are struck directly or hit by ground current. Their anchor slips and they fall about 100 feet, their drop halted only because their rope has wrapped around rocks. All three have severe burns, two lose consciousness, and one has a broken shoulder. Jake suffers a concussion.

*B*y 4:28 p.m., one of the rangers' Bell 206-L4 helicopters, nicknamed Two Lima Mike for its call letters (2LM), lands at the Lupine Meadows heliport. The pilot, Laurence Perry, had been putting out a small fire south of the nearby town of Jackson and was grounded for about ten minutes due to lightning before responding.

Perry has the suave, debonair look of a Cary Grant and speaks with a native British accent, his speech peppered with expletives. His life has been the stuff of an adventure story— he's "laughed at danger," hauling police into the jungles of Papua New Guinea, oil workers into Yemen and the Sudan, rescuing climbers from the peaks of mountains in Argentina. He has flown helicopters for thirty years. "I miss the weddings, the funerals, and the bar mitzvahs," he says. "But I enjoy the hell out of flying."

Perry is briefed by Brandon Torres, and his helicopter picks up two more rangers and whisks them up the mountain for a reconnaissance flight. They soar toward Rod, who is still hanging off the side of the cliff. As the helicopter hovers above, the rangers assume he is dead. Perry remarks on Rod's seeming insignificance surrounded by the massive mountains. Referring to the notorious alpine peak where one climber dangled for days before his death, Perry thinks it looks "Eiger-like. It was something you'd expect to see back in the old days—horror stories from the Alps."

But as the helicopter moves in closer, Perry notices Rod's right arm moving. "The guy's alive," he shouts.

The copter circles above Clinton and Erica, who is not moving at all. The rangers take photos of the three bloody and confused climbers of the fourth team, who are lying at the bottom of Friction Pitch.

Two Lima Mike lands at a rudimentary helipad that the rangers maintain at the Lupine Meadows rescue station and sends their photos to Torres, who displays them on a computer screen. He pauses at the photos of Rod. The ranger fears that the blood supply to Rod's legs might be cut off; that his harness might be cutting into his diaphragm, making breathing difficult; and that his tongue might have rolled back, blocking his airway.

"That guy's gonna die," Torres thinks to himself. Looking at the photos, he realizes that there are two accident scenes, hundreds of feet apart, which will require two separate rescue efforts.

Torres glances at his watch. It's almost 5 o'clock. Helicopters are generally not permitted to fly in the Tetons more than half an hour after sundown, the so-called pumpkin hour. Torres checks the precise time of the sunset, and adds thirty minutes—9:23, or 21:23 in the military time used by the rangers. The rangers have four and a half hours to get to Rod, recover the Summerses on the ridge above, and rescue the three injured climbers below.

Using a technique known as short haul, the rescue team plans to suspend rangers a hundred feet below the helicopter by two finger-thin strands of nylon rope, insert them on the mountain to pick up the injured, then transfer them to a second helicopter at the Lower Saddle, the nearest appropriate landing zone. From there, the injured will be taken to Lupine Meadows, where they will be triaged and then either taken by helicopter or by ambulance to a hospital. The short-haul method minimizes a helicopter's hover time and thereby decreases the risk to the rescuer.

Perry is the pilot. Renny Jackson, who wrote a climbers' guide to the Tetons and has scaled Mount Everest twice, is his spotter. Leo Larson, then a twenty-seven-year veteran ranger, standing six foot five and with a long blond ponytail, secures the nylon rope to his harness. The rope is laid out at his feet in an elongated *S*.

"Hooked and ready," Larson shouts into the radio to Perry, over the whir of the copter's rotors. The afternoon shadows are lengthening as the helicopter climbs into the sky through gusts of over twenty-five miles an hour, with Larson suspended below, defying the laws of nature in the closest possible approximation of a human flying through the air. As Perry climbs, the helicopter rises and falls in the turbulence like a roller coaster. Larson bounces up and down, the cold air stinging his face.

The solid rock walls of the Grand move frighteningly closer, but Perry is confident and relaxed. But as the copter approaches Friction Pitch, Perry mutters to Jackson, "Bloody hell." Clouds are forming, with the possibility that he will lose sight of the mountain like a skier in a whiteout, with no idea how close he is to the slope. He needs a vertical reference to navigate. A break in the clouds gives him a momentary glimpse of Rod dangling below. Then he is obscured again, and the peak is shrouded.

"Leo, we're going to have to abort," Perry radios to Larson, hanging below the copter. "This isn't going to work." Incoming clouds and a strong wind prevent the helicopter from hovering, and it is impossible for Perry to determine the condition of the climbers. Perry decreases the pitch of the blades and the chopper begins a slow, spiraling descent. He has lost sight of Rod.

*S*uspended in air, Rod hears distant voices. Somebody on a radio somewhere is saying that one and possibly two people are dead. He realizes they are talking about him. He tries to scream for help, but no sound comes. The pain is excruciating. He is having trouble breathing. He decides to reduce the weight on his back. With his good hand, he unties his pack. It crashes far below on the rocks. But by dropping the weight, he's also lost his supply of food and water, and the pain doesn't go away. *Maybe if I stop breathing*, he thinks, *the pain will lessen. Maybe if I stop breathing.* He fades in and out of consciousness.

He hears the helicopter and yells as loudly as he can. "Get me outta here," then the *chop chop* sound of the rotors quickly disappears, and the sick feeling returns.

Another climber is shouting. "Hold on, buddy. Just keep breathing. Hold on. We're going to try to lower you." Rod can hear him but can't see him. It's getting colder and the pain is getting worse. The face of his newborn son, Kai, drifts into his head. He keeps thinking about Kai and his wife, Jody.

He prays to "whomever. Please don't let me die. Please don't let me die. Not now and not yet."

*A*t 5:21, Perry returns the helicopter to the Lower Saddle. Jackson jumps off. The clouds may not permit another flight. "We've got to get people up there on foot," Jackson says to the rangers around him.

Jim Springer has loaded over thirty pounds of ropes and other rescue equipment into his backpack. Forty-eight years old, he grew up climbing in the mountains of Washington State, where he was a teenage Explorer Scout. During the winters, he is on skis practically every day as an avalanche forester, warning skiers of possible avalanches and performing rescues. *What in the world is going on up there?* Springer wonders as he heads up the mountain.

"You've got to catch Springer," Jackson shouts at ranger Jack McConnell.

"Cool," says McConnell. Known as Jack Hammer or Hydraulic Jack, for his pile-driver legs, McConnell rushes off to join Springer in the rescue effort. McConnell's best time for the 2,170-foot climb from the Lower Saddle to the summit, a climb that takes most people at least eight hours, is fifty-five minutes. This time he and Springer only have to cover about 1,200 feet to Friction Pitch, but it's tough climbing with a loaded backpack. Springer yells to McConnell, "We can beat the guys in the helicopter."

*T*he two teams of climbers above Clinton and Erica have descended to Friction Pitch and start building a rock enclosure to block the wind as the temperature drops to near freezing. They cover the shivering Clinton. His leg is bloody and purple. His running shorts are shredded. He just wants to get off the mountain, to be with his two children, Addison, four years old, and Daxton, two. The climbers have stopped giving CPR to Erica.

Brandon Torres, the rescue coordinator, gets a cell-phone call from Robert Thomas's wife, on the mountain. "We really need your help," she says, sounding exhausted. "We've stopped CPR. Erica's dead."

*A*t six o'clock, Friction Pitch is visible again. Perry can go back up. With only a little over three hours remaining, Perry's helicopter approaches the ledge, with Larson once again at the end of the short-haul rope. "I'm heading for that flat rock," Perry radios to Larson, moving toward a slab the size of a large bed. Larson's legs are twenty feet above the mountain. The chopper eases Larson lower.

From the cabin, Perry thinks that Larson looks "like a wee man on a long string." He has practiced this maneuver many times, by landing a ranger at the end of a rope on a picnic table. Larson remembers his own rescue about twenty-five years earlier, when he was injured after a rock-fall. It had taken thirty-six hours from the time his injury was radioed in until he reached the hospital, as rescue techniques were not as advanced. Now Perry eases Larson onto the rock slab.

Two hours and twenty-three minutes after the initial call for help, the first rescuer arrives on the scene. Larson checks Erica's pulse. Nothing. Then he determines the extent of the injuries to the others—who can climb down the mountain and who needs to be pulled out by helicopter. After the strike, the four climbers in the team above had rappelled down to the ledge and performed CPR on Erica for thirty to forty-five minutes before Clinton asked them to stop, urging them to help anyone else who is injured but still alive. Larson concludes that everyone in this group seems okay.

But Clinton is unable to walk. He complains of right leg and right foot pain from the lightning strike. He tells Larson that he was sitting next to his wife when she took a direct hit.

Larson points down to Rod, asking, "Is he alive?" They tell him that they've heard Rod groan.

With Larson on the ground, Perry's chopper flies off to pick up the other rangers. When he returns, five more are inserted in three trips to the ridge. Larson asks the uninjured climbers to leave the narrow ledge to give the rangers room to work, and the rangers begin to rappel down the rock face. The bright orange and yellow colors of their clothes and helmets stand out in stark contrast to the brown and gray rocks of the mountain and white patches of snow.

The rangers decide that the injured climbers of team four below also need their help. They rappel down about 200 feet to reach them, passing within fifteen feet of where Rod is suspended. "Hang in there," one ranger says. "People up above are setting up ropes." There is no response.

Meanwhile, the two rangers who practically sprinted up the mountain, Springer and McConnell, reached the spot where the fourth team is stranded. It took them only a little over an hour from their base. They arrive just as the ranger who is rappelling down the rocks appears. They have to keep the three injured climbers alert and awake. To sleep may be to die.

One of the climbers has open bilateral leg fractures that are still bleeding. Another has a leg injury and burns. A third has rib injuries and burns and is drifting in and out of consciousness. "There was a tangle of ropes and a lot of blood," Springer recalls. "They had tumbled down about sixty feet before their rope snagged. It's incredible they weren't hurt more."

The rangers conclude that all three can be evacuated off the mountain in a "screamer suit," a nylon vest with a large diaperlike bottom attached to the helicopter's hundred-foot-long rope.

The unhurt climbers from team one and team two are climbing down the mountain, while another helicopter delivers a slingload of rescue equipment and sleeping bags to the upper scene, where Clinton and Erica Summers lie inert. The early-night air is turning colder.

At seven p.m., Perry's helicopter takes off from Lupine Meadows with an empty screamer suit slung below and arrives at Friction Pitch a few minutes later. Jackson, the spotter, stares down between the skid and the fuselage. The rope has to be positioned exactly to reach the rescuers. Perry pokes his head out, rocks the machine slightly, and maneuvers the carabiner into a ranger's palm. It is exceptional piloting.

With the helicopter hovering above Friction Pitch, the rangers strap Clinton Summers into the screamer suit. With serious burns and other injuries, he is the first to be lifted off the ridge. At the Lower Saddle, more rangers are waiting. They remove the extraction harness, quickly check Clinton's condition, and load him onto an identical helicopter, which carries him down to Lupine Meadows. There a Teton County ambulance will wait with him for the other injured victims before rushing them to St. John's Hospital in Jackson, Wyoming.

Two of the climbers from team four who were wedged

National Park Service rangers rescue an injured climber be-
lieved to be Rod Liberal from Friction Pitch on the Grand
Teton after he was struck by lightning on July 26, 2003.
(Photo by Leo Larson, National Park Service)

into a narrow groove in the mountain are still tied up in
their ropes. As one climber is placed in the helicopter's sling,
two rescuers stand ready with their knives to make sure that
the ropes don't tie the helicopter down. As the climber is set
to be evacuated, a ranger tells him, "Get ready for the best
ride in the amusement park." Bob Thomas, who made the
first emergency cell-phone call to the rangers and then
climbed down Friction Pitch to aid the injured, is next, followed by another climber. The last to leave is Rod's friend
Jake Bancroft, who flashes a V-for-victory sign as he ascends
into the sky.

*T*o Rod, though it isn't yet night, everything is black. He
simply can't see anymore. He hears a helicopter, the sound
of its rotors becoming louder and louder. Finally it seems as
though it's right on top of him. Then he loses consciousness
again, thinking that it's finally his time to "check out."

He's still unconscious when Ranger Craig Holm finds
him. When Holm surveys what's happened, he thinks "it
was the wildest scene I had ever been on. It looked like a
bomb went off."

Steadying himself on a ledge under the weight of his
forty-pound backpack, Holm peers over to where Rod has
been dangling upside down in the air. He has been held by
only his rope for over four and a half hours—as if the fates
cannot decide his destiny.

The ranger shouts: "Rod? Rod?" But there is no answer.

At thirty-five, Holm has thinning sandy hair and a youthful, choirboy face. He grew up in Monterey, California, and

after college moved to Boulder, Colorado, where he worked as a climbing guide and a National Park ranger with special medical training.

The ranger anchors himself to the mountain, rappels slowly down forty feet, and stops, hanging near Rod's waist. He attaches a tether from his harness to Rod's belt in case the rope that is holding the injured climber was damaged by the lightning, as it takes a fairly low temperature to damage nylon.

"I'm Craig," the ranger says. "We're here to get you out.... How are you doing? Talk to me."

Rod mumbles incoherently.

Holm asks the four questions rescuers use to determine the mental state of an injured person: What's your name? Do you know where you are? Do you know what today is? Do you know what happened?

Rod doesn't seem to know anything except his name. He is still doubled over in a position that Holm recalls "didn't look humanly possible without breaking your back." Rod passes in and out of consciousness, his breathing a kind of snoring.

Holm takes Rod's pulse and puts an oxygen mask over his mouth and nose. He braces his legs against the rock wall at a ninety-degree angle, then fashions a chest harness from tubular nylon slings and wraps it under Rod's arms and around his back, lifting him and placing Rod on his lap. Now both are facing the wall, and Rod's breathing improves. At five foot ten and weighing 140 pounds, Holm is only about ten pounds lighter than Rod, but the injured climber feels much heavier—he is deadweight.

Rod asks, "Am I okay?"

Holm replies, "What's the name of your son?"

"Kai," comes an answer from somewhere deep inside Rod's chest.

"How old is he?"

Again, so faint it's less than a whisper: "Three months." Rod is too weak to tell Holm about the pain.

Holm assesses Rod's condition. He unzips Rod's jacket and pulls up his T-shirt. Under Rod's left arm, across part of his chest, is a spidery purplish six-inch mark. Rod winces when Holm presses his lower back, hips, and left side—his chest, arms, and shoulders. "Does this hurt?" he asks, pressing Rod's ribs. A moan. Rod can squeeze Holm's fingers with his right hand but not his left. His clothing has fused to his body.

The sun is lowering over the peaks. The two are suspended side by side next to the solid granite wall. Rod continues to breathe shallowly, and he's freezing. Meanwhile, one by one the other climbers are attached to the screamer suit and lifted off the mountain to safety.

Holm wants to put a down jacket on Rod, but an oxygen tank is stuffed into Holm's backpack. Cradling Rod's head with one hand, Holm can't safely pull the jacket out. If the oxygen tank slips, it could fall on the climbers and rescuers below.

"George!" Holm calls up the mountain. "I'm going to need an extra hand down here."

Ranger George Montopoli, who had been setting anchors above, is one of those people who seem to excel at everything. Originally from New York State, he entered college at the University of Colorado in Boulder and began climbing. Since then, he has climbed mountains in Ecuador,

Chile, and Argentina. A friendly man with salt-and-pepper hair and bushy eyebrows, Montopoli served in the Peace Corps, is a bald-eagle researcher, and has a Ph.D. in mathematics. "My strong point as a rescuer is my medical experience from the Peace Corps," he says.

As Montopoli descends, Holm keeps talking to Rod. Holm figures Rod has only a fifty–fifty chance, but says nothing. Clouds are gathering but the wind has subsided. It is cold: The temperature has fallen into the forties. Rod is shivering. Holm regularly checks his pulse and his breathing. The scene is eerily calm.

After rappelling down the sheer wall, Montopoli reaches into Holm's pack, moves the oxygen—Holm is giving Rod oxygen from another tank—and hands him the down jacket. "Holm was talking to him," Montopoli recalls. "We didn't think he would survive the night. He was really hurt. He took the full blast of the lightning."

Now that Rod has been stabilized, the rangers have to figure out how to get him off the mountain. Because of his injuries, he can't be strapped into a harness like the others but will have to be taken out in a plastic basketlike litter suspended from the helicopter.

The rangers above lower down the orange fifty-pound basket. While Holm holds the litter as steady as possible, Montopoli crouches like a cat on its narrow rims. They pull the backboard halfway out and gingerly ease Rod onto it. Then they carefully slide him into the basket. With Velcro straps, they secure his shoulders, chest, and legs and place his head between two spongy blocks. For warmth against the evening cold, they bundle him in a sleeping bag.

"Okay," Leo Larson says from the ridge above. "Let's go."

Montopoli returns to the ledge above, and he, Larson, and two other rangers start to raise the litter with a mechanical pulley. Foot by foot, Holm walks up the side of the rock wall in an L-shape position, steadying the litter at his waist to prevent it from hitting the rocks. As they ascend, Holm keeps talking to Rod. "We're gonna get you out of here. It's almost over. Tell me about your son."

It takes thirty minutes for the litter to be raised to the ledge above Friction Pitch. Rod is still lying on his back, but the rescuers want to repackage their precious cargo to place him on his side, so he won't choke if he vomits. Holm glances at his watch. It's 8:55—two minutes past sunset. There isn't enough time to strap Rod in another way. Instead, the decision is made to fly Rod "attended"—so an attendant can manage Rod's airway should he vomit. They will fly all the way to Lupine Meadows instead of to the Lower Saddle, the nearest landing zone, an eight-minute instead of a one- or two-minute flight.

Holm is clipped to the rope. The load now weighs over three hundred pounds, including the basket, Rod, and Holm—an awful lot of cargo for a helicopter to haul at 13,000 feet. Holm will hang alongside Rod in the litter as the helicopter descends all the way down to Lupine Meadows, some 6,300 vertical feet below.

UNEXPECTED CONSEQUENCES

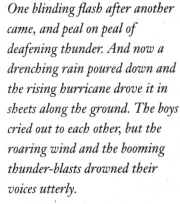

One blinding flash after another came, and peal on peal of deafening thunder. And now a drenching rain poured down and the rising hurricane drove it in sheets along the ground. The boys cried out to each other, but the roaring wind and the booming thunder-blasts drowned their voices utterly.

—MARK TWAIN, *THE ADVENTURES OF TOM SAWYER*

r. Michael Cherington, whose youthful, rosy face belies his seventy-seven years, first became intrigued by lightning in the early 1990s after treating a patient who had been struck while riding a bicycle. An avid cyclist himself and a founder of the Lightning Data Center of St. Anthony Hospital in Denver, Colorado, Cherington was curious to learn how susceptible cyclists were to lightning. Research eventually showed that golfers, fishermen, beach and boating enthusiasts, and other outdoor-sports participants, including bicycle riders and motorcyclists, were at the top of the list.

In the United States, you are statistically most likely to

be hit by lightning on a weekend summer afternoon if you are standing under trees, boating, participating in other water activities, or playing golf. The greatest risk is in "Lightning Alley," the corridor from Tampa on Florida's west coast to Titusville on its east coast.

The golfing casualties occur when players seek safety under trees or beneath ungrounded structures or while playing the game itself. An extreme example of a golfing accident occurred in the late spring of 2004, when nineteen golfers in Kremmling, Colorado, were hit by a single flash of lightning. They were participating in the Kremmling Cliff Classic—a "tournament" that involves hitting golf balls off the edge of the cliff at targets in the valley below—when one of the golfers' clubs was hit by lightning. The flash then jumped from one person to another. Fifteen golfers suffered minor injuries, and four were taken by helicopter to a hospital in Denver for life-threatening injuries, including serious burns, although all have since recovered.

The threat of a lightning strike is so prevalent among golfers that golfing great Sam Snead once remarked, "The three things I fear most in golf are lightning, Ben Hogan, and a downhill putt." Champion golfer Lee Trevino, who was knocked unconscious after a lightning strike, quipped, "If you are caught on a golf course during a storm and are afraid of lightning, hold up a one iron. Not even God can hit a one iron."

Although in many countries there is no information on the number of lightning deaths, lightning statistician Ronald L. Holle estimates that every year 24,000 people are killed from lightning worldwide and 240,000 are injured. He assumes that the susceptibility to lightning in many regions of the world is similar to the rate that applied in more-developed countries in 1900, when most people lived in rural areas. At that time, the rate was about six lightning deaths per million yearly, with a ratio of ten injuries for every death. He bases current figures on a population of four billion people who live in the tropical and subtropical regions of the world, where there is a high degree of lightning.

In the developed countries, the rates are much lower. For example, annual fatalities during the latter part of the twentieth century ranged in France from five to thirty-four; in Japan, three to fifty-eight; in England and Wales, zero to eleven; in Germany, three to seven.

In the U.S., lightning fatalities have declined from about a hundred annual deaths from 1959 to 1991 to between forty and sixty deaths and between 350 and 550 injuries, also a decline. There are a number of reasons for the decrease, including a greater awareness of the threat of lightning, improved meteorological forecasts, more substantial buildings, a decline in the rural population, and improved detection methods.

Inexpensive handheld detectors are available, although there is anecdotal evidence that they often do not locate lightning accurately or detect weak or infrequent bolts, which can still be deadly. Professional-grade lightning-detection systems are available that use electromagnetic sensors in a number of locations and generally perform well. Lightning-detection services can be purchased on a subscription basis. They provide automatic notification, which can be sent via pager, e-mail, or cell phone, when lightning has been detected within a given range by the National

Lightning Detection Network, which can detect flashes from about a hundred to over a thousand miles away. Similar sensor networks operate in a number of countries around the world. Golf courses have installed systems that use sirens to alert golfers when lightning strikes are nearby.

Each June, National Lightning Safety Awareness Week is sponsored by a number of concerned organizations, including the National Oceanic and Atmospheric Administration (NOAA), the National Weather Service, Little League, the PGA golf tour, and Struckbylightning.org, an advocacy group for lightning safety and education, which maintains a database of lightning strikes worldwide. It was founded by survivor Michael Utley. Among its catchy slogans are: *Don't be lame! End the game! Don't be a fool! Get out of the pool!*

Clearly, being caught near water poses dangers. The leading locations for lightning deaths and injuries are people standing under trees and in or near water, according to the National Weather Service. The most dangerous water activities are fishing, swimming near beaches or being on a beach, and boating.

A number of incidents have been recounted of fishermen struck standing in or near water. For example, in September 1999, Dave Grillmeier was fishing with a friend at a lake in Orlando, Florida. As a storm approached, he remembers he "grabbed his graphite rod and walked back . . . to get one more cast before leaving." Just as he "cast his worm," lightning hit him. He spent thirty days in the hospital.

Obviously the open spaces of beaches lend themselves to lightning strikes. When a bolt struck New York's Neponsit Beach in 1994, one person was killed and seven were injured. Of course, anywhere in a pool area is also vulnerable to lightning.

Ever since humans began sailing, lightning has posed a serious threat. There are a number of hair-raising accounts of ships being struck. A particular threat was lightning hitting the powder magazine on naval vessels, causing an instant explosion. Benjamin Franklin was never completely able to solve the problem of strikes on ships. But a solution was found by William Snow Harris, a physician and lightning-protection advocate, who suggested that the ships' masts themselves be turned into lightning rods by lining them with copper plates that could be connected to the hull and then to a chain into the sea. He wrote publicly about his invention in 1820, but it wasn't adopted by the Royal Navy until 1847 because of bureaucratic hesitation.

Metal ships are rarely damaged. They are frequently struck, but the high conductivity of their large quantities of metal, with hundreds of square yards of hull in direct contact with the water, causes a rapid dissipation of electrical charges. But small boats are seldom made of metal. Their wood-and-fiberglass construction does not provide the automatic-grounding protection offered by metal-hulled craft. Fiberglass boats, especially sailboats, are particularly vulnerable to lightning strikes, since any projection above the flat surface of the water acts as a potential lightning rod. In many cases, the small-boat operator or casual weekend sailor is not aware of this vulnerability to the hazards of lightning. But these boats can be protected from lightning strikes by properly designed and connected systems of lightning protection, similar to those installed on large ships.

One danger on open water is from side splashes of lightning. For example, in the summer of 2007, a man and his nine-year-old son were out in their small boat near Crab

Island in the Florida Panhandle when a bolt of lightning struck the water, knocking the father unconscious and blinding him for several minutes. Fortunately, his son was able to pilot the boat to shore.

In water, the lethal radius from a bolt's point of contact is about six hundred feet, compared to sixty to a hundred feet on land. With such a risk, swimmers should leave the water when there is any chance of lightning.

Outdoor group sports with a risk of lightning include soccer, baseball, and football. (Basketball is played either indoors or near shelters.) One of the worst soccer-related lightning accidents occurred in 1995 in Honduras, when seventeen people were killed and thirty-five injured as a crowd stood under a shelter next to a game during a heavy rain. A prominent American accident occurred on a Saturday afternoon in 1980 in southern Illinois, near St. Louis, when children, mainly ten- to thirteen-year-olds, returned to the soccer field after spending an extended halftime in their parents' cars due to a thunderstorm. Shortly into the second half of the game, a bolt to the center of the field knocked down all the players and most of the children and adults on the sidelines. Six children required immediate medical attention and one boy was killed.

Holle has identified twenty-eight baseball and softball events in which people were killed or injured, and most of those struck were teenagers. Safety experts advise that games be halted until the threat of lightning passes and that adequate shelters be provided nearby for players and spectators.

In what may have been the most disastrous lightning accident on a football field, a high-school player died and about forty players and coaches were injured after a bolt struck during practice at Grapeland High School, East Texas, in September 2004.

*O*nly after instructions on the use of sumps (a hole in the ground or other wastewater depositories); the proper disposal of trash; the cleaning of latrines; and the prevention of dehydration, heatstroke, hyperventilation, hypothermia, sunburn, and acute mountain sickness does a recent, official Scout Jamboree manual get around to the topic of *Lightning and Flash Floods.* The danger to campers and hikers from lightning is best illustrated by accidents that have occurred to Boy Scouts. Perhaps no other group has had such a complicated relationship with lightning as the Boy Scouts of America.

In July 2001, Mark Evans, a sixteen-year-old Eagle Scout, was returning from the activities center to his campsite at the National Scout Jamboree at Fort A. P. Hill in Virginia. As a storm threatened, Mark remarked to a friend, "Wouldn't it be weird if one of us got struck by lightning?" Minutes later a shaft of light arced through the air, slamming a number of the boys to the ground.

All the scouts quickly got up, except for Mark. He wasn't breathing and his heart had stopped. His scoutmaster immediately performed CPR but still could find no pulse.

Mark and his group were all members of the Church of Jesus Christ of Latter-day Saints (Mormons). When Mark failed to respond, another scoutmaster bent over him, uttered a prayer, and sprinkled consecrated oil on him. After several minutes, Mark began to breathe and regained consciousness. "What happened?" he asked. "Did you get it on film?" No one had filmed Mark's accident, but the lightning bolt had reset the calendar on the scoutmaster's watch back to the manufacturer's original date.

Now twenty-two years old, Mark enjoys good health and lives in California. After leaving the hospital the morning after the lightning strike, he returned to the jamboree and went on to earn several merit badges in his scouting career. Since then, he has not suffered any neurological or other health problems. A cardiologist later told him that the chances of avoiding serious complications after five minutes of cardiac arrest are very low. "I believe that because of that scoutmaster's blessing," he says, "I was able to be revived."

Mark was one of the lucky ones. From 1995 through 2005, seven scouts and scout leaders were killed and about fifty were injured in fifteen lightning incidents at scout camps or on scouting expeditions. Given the disproportionate number of casualties, a number of lightning authorities have been critical of the scouts' safety record. "I was an Eagle Scout—I love the Boy Scouts," says Air Force meteorologist William Roeder. "But on this one they're wrong. I do believe that the Boy Scouts of America need to improve their lightning safety."

In August 2002, sixteen-year-old scout Matthew Tresca was killed in another high-profile incident. Matthew had been participating, along with more than three hundred other scouts, at a summer camp in the Pocono Mountains near East Stroudsburg, Pennsylvania. As the week was drawing to a close, with lightning flashing in the distance, scout leaders sent the boys back after dinner from the dining hall to their tents in the woods. At about seven p.m., lightning struck a tent pole near a picnic table where Matthew was sitting under a tarp. The teenager suffered cardiac arrest and was pronounced dead ninety minutes later. After the incident, Matthew's parents brought a lawsuit against the Boy Scouts of America, alleging that proper training and planning could have prevented Matthew's death.

At the trial, Robert Shaker, a parent and Tresca's scoutmaster, testified to the chaos at the campsite when the lightning bolt struck. He had been sitting on a picnic bench under a tarp talking to the father of another boy, when a blinding flash and deafening explosion rocked the campsite.

"I saw a lot of people on the ground," Shaker testified. He recalled that a seven-year-old girl visiting her brother had been knocked down and was screaming that she didn't want to die. Shaker's son was also thrown to the ground, crying and screaming that he couldn't feel his legs. Tresca had been knocked over backward by the bolt. "His eyes were open," Shaker told the court. "He was staring straight up. I immediately shook him, trying to get a response from him. I shook him and shook him. He just had that blank stare."

Another camp employee testified that prior to the lightning strike, he had noted the darkening skies and increasing wind with some concern. "It's something that I'd only felt a few times before," he said. "It seemed to be a charge in the air, a palpable electricity, the kind of feeling that makes the hair on your arms stand up." He had recommended to the camp leader that the scouts remain inside until the bad weather had passed.

For their part, Boy Scout officials denied any negligence, describing Matthew's death as "an act of God" and blaming a "rogue bolt of lightning" from skies that seemed to be clearing. But Ronald Holle wrote in a court brief submitted on behalf of the Trescas that Matthew's death had been preventable. "If only on the basis of hearing thunder and seeing the flashes during the day," he argued, "trained people should have kept everyone in the dining hall." Before the Trescas and the Boy Scouts agreed to a confidential settlement in 2006, Mary Tresca, Matthew's mother, said of the tragedy, "If it's going to take the Boy Scouts getting hit in the pocket to protect anyone else's family, then that's what I guess it takes."

In fact, it may have cost the Scouts even more. In a case that was recently settled, the family of James Rozwood filed a multimillion-dollar claim against the Boy Scouts of America and the Iroquois Trail Council, a local scout group in western New York, in connection with a July 2001 lightning strike at a scout camp rifle range that left Rozwood, then sixteen, a quadriplegic who can communicate only by blinking his eyes. Like the Trescas, the Rozwoods alleged that the Boy Scouts lacked an effective lightning-safety plan and disregarded the warning signs of foul weather.

In the summer of 2005, there were two more fatal lightning accidents involving scouts. In one incident in August, a group of scouts were camping in Utah's Uinta Mountains when lightning struck a tree and the current passed through the metal nails in a log shelter in which the scouts were sleeping. One boy was killed by the shock and three others were injured. In another case, a number of scouts were following the John Muir Trail on a nine-day hike in July to Mount Whitney in California. Lightning had been flashing throughout the afternoon (according to data ground sensors operated by the National Lightning Detection Network, about 416 strikes had hit the ground within fifteen miles of the mountain). As they huddled under a tarp strung between the trees, a bolt of lightning flashed, killing one scout and a scout leader.

After the disaster, a ranger and spokeswoman for Sequoia National Park commented, "The only thing they could have done differently was simply disperse a little bit more, but actually they did as well as they could do in the situation that they were in." In a textbook scenario, the scouts might have been better protected if they had positioned themselves farther apart, in squatting positions, but Holle says that critiquing what the scouts should or shouldn't have done in that situation is beside the point. The safest response would have been to head down the mountain immediately or, better yet, not to have been on the mountain at all. "My recommendation is—just as you don't go up on Mount Whitney on the third of January in a raging snowstorm, you don't go up in raging thunderstorms at certain times of the year."

Naturally, the number of tragic incidents involving scouts has been troubling for the boys and their parents, as well as the organization. "We take lightning and all safety issues extremely seriously," said Gregg Shields, a spokesman for the Boy Scouts of America, "and we try not only to protect our people as much as possible but educate them to enable themselves to be safer for the rest of their lives as a result of their scouting experience. We try to educate everyone about the dangers of lightning, the very best ways to minimize the chance of being struck by lightning, and the best way to treat people in the event of an unfortunate accident." The National Council now encourages and offers its local councils access to a lightning-alert service.

*E*ven for those who are engaged in more-passive leisure activities, lightning can pose a dangerous threat. It has long been known that the current from lightning can travel through land-based telephone lines during electrical storms and cause serious injury. Recently, more portable electrical devices have attracted suspicion as possible conductors. Though many experts are skeptical, there are numerous reported cases of individuals who have been struck while carrying cell phones, leading to burst eardrums, cardiac arrest, and residual cognitive problems.

iPods may pose an even greater threat. One Sunday afternoon in 2006, seventeen-year-old Jason Bunch of Castle Rock, Colorado, was mowing his family's lawn. He was listening to Metallica on his iPod one minute and the next he found himself in bed with burns all over his face, vomiting, and bleeding from his ears. Although it's not clear whether Jason's iPod acted as an antenna for the bolt of lightning that ran through his body, the damaged earphones and a hole in the back of the iPod's case would seem to indicate that it did serve as an electrical pathway. He suffered burns from the earphone wires on the sides of his face, a burn on his hip where he was carrying the iPod in a pocket, and a burn up the side of his body. He was knocked unconscious and lost hearing in one ear and his sense of equilibrium. Ironically, the earphone cables may have directed the current away from his chest and, crucially, his heart.

The New England Journal of Medicine reports another incident in which a thirty-seven-year-old jogger was struck by lightning while standing under a tree in Vancouver, British Columbia, during a thunderstorm in June 2005. An active church musician, he was reportedly listening to religious music on his iPod at the time. His injuries included a pattern of burns on his chest along the path of the earphone wires. "The victim had earphones on and had been sweating from jogging . . . and the earphones transmitted the electrical current into his head," according to Dr. Eric Herrman, a radiologist at Vancouver General Hospital. The jolt, which also broke the victim's jaw in four places, left him with damaged tympanic membranes (eardrums), dislocated bones in the middle ear, and a fifty percent hearing loss. But lightning medical experts are skeptical that iPods attract lightning.

"There is no evidence that a metal or electronic apparatus worn or carried on the body, whether on the head or elsewhere, makes a person more attractive to lightning," specialists Dr. Mary Ann Cooper and Dr. Chris Andrews wrote in a recent letter to the *New England Journal of Medicine*. "Eardrum perforation is the norm in lightning-related injury, not a sign of any special effect due to an iPod."

Despite the best precautions, the nature of weather is to be unpredictable, and the odds are good that at one time or another we'll all be overtaken by an unforeseen thunderstorm or a violent change in the weather. In such cases, lightning experts have devised a number of basic safety rules.

(1) If you're planning an outdoor activity, check the local weather forecast for any predictions of rain and thunderstorms.

(2) Employ the 30–30 rule: After you see lightning, count the time until you hear thunder. If it's thirty seconds or less, go immediately to a safer place. After the storm has apparently dissipated or moved on, wait at least thirty minutes before leaving the safer location.
(3) When lightning threatens, go to a safer location. What is a safer location? A fully enclosed, substantially constructed building. Once inside, stay away from telephones and appliances, electric sockets, and plumbing, as the current can be carried through electrical wiring and metal pipes. Don't stand in an open doorway or near an open window.
(4) If you can't reach a building, an enclosed vehicle with a solid metal roof is a good second choice, but close the windows.

As cyclist Dr. Cherington of Denver's Lightning Data Center notes, the enormous force of lightning is released on defenseless people who are often engaged in outdoor recreation and sports events. "Many, if not most, of these cases could have been prevented."

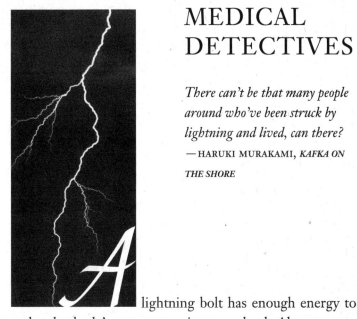

MEDICAL DETECTIVES

There can't be that many people around who've been struck by lightning and lived, can there?
— HARUKI MURAKAMI, *KAFKA ON THE SHORE*

A lightning bolt has enough energy to make the body's systems go into overload. Almost every bodily system is vulnerable, and a wide variety of complications can result. Lightning causes fatal injury to only about nine to ten percent of those struck. But disability from a strike can be a serious problem. While the effects of a lightning strike can be devastating, the relatively small number of victims means that most doctors and hospitals see few survivors, if any, and when they do, they often don't know how to treat them.

Tony Scott, a sixty-two-year-old retired telephone repairman in Hollywood, Florida, recalls how he was struck by lightning on July 29, 1990, while installing a new line for a customer. It was a clear day, no rain in sight. Suddenly, "There was a bolt out of the blue," he says. "It hit the pole and traveled down the line. I was thrown across the yard fifteen or twenty feet. When I came to, I was paralyzed."

Three years later he was waiting in a customer's driveway when lightning hit a nearby tree. "I didn't have any medical problems beforehand," he says. "Since then I've had heart problems and diabetes. I ended up having quadruple-bypass surgery, and I've had eight balloon procedures. The possible side effects from lightning are the things that most doctors don't realize. You've got chronic fatigue and chronic pain. They can't find anything wrong with you, and they tell you it's all in your head. Go see a psychiatrist," he remembers being told, "which adds to the problem, 'cause you go around thinking you're crazy."

Another victim, Brad Anderson, who was struck on his dairy farm near Buffalo, New York, in 1977, recalls that in the emergency room the next day, "The doctor there said nothing was wrong, take these pills and go home." Though he continued to suffer from chest pains, headaches, and shaking in his arms and legs, and despite repeated tests—CAT scans, X-rays, blood work—nobody could ever determine what was wrong with him. One doctor accused him of faking his symptoms and told him to get out of his office.

Lisa Hall has endured an especially long medical odyssey after she was struck by lightning in her living room in West Virginia in 1977, when she was fifteen. Right after the strike, she says, she "felt fine" and didn't seek medical treatment. But after she was hospitalized for genetic ulcerative colitis (a disease of the large intestine) in 1995, she began to display symptoms of neurological problems that several doctors later concluded may have originated from the lightning strike. She felt dizzy and had problems walking. "There was a strange hissing sound in my ears," she recalls, "and my vision became dark, with a strobelike flashing sensation." Since 1995, she has met with thirty-eight doctors, including neurologists; psychiatrists; neuropsychiatrists; ear, nose, and throat specialists; chiropractors; acupuncturists; internists; naturopaths; even psychics. She traveled to hospitals at the Mayo Clinic, Johns Hopkins University, Vanderbilt University, and the University of Alabama at Birmingham, among others. "I was diagnosed with everything from low blood volume to chemical imbalance, to deep inner-ear trouble to organic mood disorder, to obsessive–compulsive disorder (based solely on the way my medical records were arranged), to suffering from a conflict about my role as a Southern woman (my personal favorite). One doctor just threw my medical binder across the room." Eventually, after finding doctors who diagnosed her condition and prescribed the correct medication, she improved greatly.

Soon after he and his wife were struck by lightning in 1998 at Greenwood Lake, New Jersey, Paul Aurand felt like he had run "a marathon and then got run over by a steamroller." He was still in shock and "sore everywhere." His insurance company told him to return to the hospital at once. "If my health-insurance company is insisting we go back to the emergency room immediately, I know we are in rough shape," he said. "But three doctors and two emergency medical clinics refused to see us. They just don't know what to do for lightning victims." Today, after treatment that included acupuncture, massage therapy, empowerment group therapy, and mineral and vitamin supplements, he feels stronger.

*T*he effects of a lightning strike differ from one person to another and are unpredictable. A person is highly unlikely to die immediately after being hit unless he suffers cardiac arrest, in which case the victim's heart, which is controlled by electrical impulses, comes to a complete stop. But lightning can cause a vast constellation of other serious injuries and problems.

When someone is struck, the current may produce a violent muscular contraction that throws the victim several feet, possibly causing fractures of the skull, ribs, and spine and dislocations of the extremities. The person may fall into a comatose or semicomatose state. The strike can affect heart function, motor control, cholesterol and blood-sugar levels, or the entire immune system, and can cause epileptic seizures. More than half of lightning victims suffer eye injuries and may develop cataracts, usually within the first few seconds but sometimes as long as two years after a strike. Temporary deafness is also common. Approximately fifty percent of lightning victims have a rupture in one or both tympanic membranes caused by the blast, especially if they were using the telephone during a thunderstorm. More than half of all lightning-injured patients experience some paralysis, though it usually disappears within minutes or days, with the affected extremities appearing cold, clammy, mottled, and insensitive to touch. But these percentages, taken from an early research paper that relied on published reports, are higher than would be expected in the general population.

It is an ancient and ongoing belief that a person who is struck by lightning bursts into flames or is reduced to a pile of ashes. In reality, lightning leaves few external signs of injury. Due to the different physical properties of lightning and electricity, a lightning shock is not like an electrical shock. If someone grabs a high-voltage line, the exposure may be prolonged because the victim often freezes to the circuit and the current enters the body. Lightning may reach energies of 100 million to a billion volts (a measure of the pressure under which electricity flows) and between 10,000 and 200,000 amperes (the unit of measurement of the rate of flow of electric current), levels seldom attained in high-voltage shocks. Lightning is a direct current, which is less dangerous than the alternating current usually associated with electric shock. But because the skin itself is a relatively poor conductor of current, most of the electricity transmitted during the milliseconds of a lightning strike flows over the body's surface. This "flashover" diminishes the force of the energy entering the body, saving the lives of most victims.

Lightning burns tend to be secondary burns. For a split second, the air in a jagged bolt of lightning can heat up to more than 50,000 degrees Fahrenheit (hotter than the gases in the sun), resulting in thermal burns if the victim's clothing is ignited and second- and third-degree burns if the victim is wearing a metal necklace or belt buckle or is carrying pocket change, which all become rapidly heated by electrical energy. Moisture on the skin can reach scalding temperatures and leave linear burns several inches long, especially in areas with heavy concentrations of sweat, such as beneath the breasts and arms or along the chest, and the superheated moisture can blast off the victim's clothing and shoes.

Lightning victims frequently experience some amnesia and confusion, regardless of whether they were knocked unconscious by the strike. These symptoms may last for hours or days. A fall resulting from a lightning strike or the concussive force from a shock wave may result in intracranial injury.

Victims are also prone to a range of symptoms, including headaches, dizziness, nausea, flashbacks, mood abnormalities, feelings of isolation and depression, heightened anxiety, hyperirritability, memory deficits, sleep disturbance, loss of libido, panic attacks, post-traumatic stress disorder, and dysphasia—difficulty with language that may include finding, understanding, or pronouncing words, or using them to express ideas. The autonomic nervous system is affected—causing hypertension and impotence as well as changes to blood pressure and cardiac response—as is the peripheral nervous system, leading to chronic pain and sensory problems.

Perhaps the most common cognitive effect of lightning is attention deficit. In this state, survivors are unable to focus attention for more than a short period of time and suffer a loss of mental agility. They are neither able to coordinate multiple tasks simultaneously nor to follow orders for complex tasks that they used to perform easily before the injury. They may have trouble coding new memories or accessing old ones. Nearly all the survivors interviewed complain of memory failure—people's names, dates, places. "I stopped to put gas in the car and wondered why the tank was already full," recalled one survivor. "My son reminded me that I just stopped a few minutes earlier and filled it up. I had no memory of doing it."

Many people with so-called mild to moderate traumatic brain injuries go undiagnosed and untreated, and they may walk, talk, and appear normal while at the same time experiencing significant problems socially and emotionally. Some victims may not trace their pain and injury directly to the lightning strike.

In the face of these persistent and inexplicable problems, physicians may doubt the patient, the family may doubt the patient, and, as attorney Charles "Nick" Simkins, who specialized in traumatic brain injuries, noted, "most sadly, even the patient may end up doubting the patient. If not properly diagnosed, treated, and respected, the patient may, and does, go on to lead a life of quiet, or not so quiet, desperation with failed employment, families, and personal lives." Yet only a handful of specialists are comfortable treating the effects of lightning on the body.

When a survivor walks into a doctor's office and says, 'I was hit by lightning,' doctors have any of several responses," observes Dr. Mary Ann Cooper, who is considered the leading authority on lightning injury, or kerauno medicine, in the world. "The physician can say, 'I don't know anything about lightning, you'd better see a specialist' or 'You're a nut, get out of my office,' or 'I don't know much about it but I'll do as much as I can to help you.' None of these inspires confidence."

Cooper is a professor in the Departments of Emergency Medicine, Neurology, and Bioengineering at the University of Illinois at Chicago, where her small office is chockablock with lightning paraphernalia and tchotchkes. A poster warning that *Lightning Kills* hangs on the door. A small action figure of Benjamin Franklin rests on a cabinet shelf above her desk. Behind her desk hangs a large photograph of a thunderbolt over Tucson, facing a poster on the opposite wall entitled *Whirlwind in Lightning.*

When Cooper was born in 1949 in Vincennes, Indiana, "only doctors' sons had the right to go to medical school," she recalls. Though her parents had been reared during the Depression and never went to college, Cooper remained undaunted and received her M.D. from Michigan State University in 1975. Years ago, a mentor told her that if she wanted to secure a reputation, she should find a niche and pursue it. "It was simple to get into lightning, because nobody else was doing it," she says. Since then, she has received numerous honors and awards for her work, including special recognition from Lightning Strike and Electric Shock Survivors International for her "wisdom, mercy, gentility, and humanity."

Wearing rimless glasses, a pink blouse, and black pants, her light-brown hair cut in a bob, Cooper is a lively speaker, changing her pitch and modulation and using her facial expressions and body movements to illustrate her points.

"It's easy to be the international authority when nobody else is," she says. "My career has been atypical, but very rewarding." She is especially proud of her fellowship in the American Meteorological Society—the only physician ever inducted—which gave her a special award "for outstanding work on the medical effects of lightning, which has enhanced the treatment of lightning-strike victims and revolutionized lightning safety worldwide." Throughout her career, her battle to educate the public about lightning injuries and safety has helped reduce the number of fatalities.

"I learned that writing for the medical literature did

very little," she says. "But going out and doing seven or eight Discovery Channel programs and eight or nine *National Geographic* programs over the last fifteen years makes a difference. It's a wonderful media story. Lightning is exciting. It's beautiful. There's science. There's heartbreak and there's recovery."

Cooper's interest in lightning goes back to her child-hood. "My father loved lightning, and he swore that he was not going to have his kids raised hiding under beds and closets like his whole family did when there was thunder and lightning. So we made parties with popcorn—whoa, fire-works! When he was older, one of his dogs used to love watching thunderstorms, so she'd go and get Dad and sit in front of this old screen door and look out and watch the thunderstorms, and if Dad didn't come and sit by her she'd go over to him and bring him over."

In the late '70s, Cooper, who was trained in emergency medicine, was giving a lecture on electrical injuries when, at the end of the lecture, a nurse held up her hand and asked, "What happens with lightning? Is it the same thing?" When Cooper researched the topic in medical textbooks, she couldn't find anything that described the prognosis.

"If someone came into the emergency room," she wondered aloud, "what was I going to tell the parents? Was their child going to live? Was their child going to die? Are they going to have permanent problems?"

In the keypunch days of the late '70s, she coded all the cases of lightning injury she could find and, after a summer of frequent thunderstorms, added her own observations to the data. Her research provided the basis for her by-now classic 1980 article, "Lightning Injuries: Prognostic Signs for Death," published in *The Annals of Emergency Medicine*. It was the first study to define the factors most likely to cause death in someone injured by lightning. A number of her correlations have continued to hold up. The only immediate cause of death by lightning is cardiac arrest (though injury suffered during a fall after the shock can lead to additional fatalities), and if you receive burns around the head you are more likely to have cardiac arrest.

Before Cooper's seminal study, very little information about lightning injuries appeared in the medical record. In the eighteenth century, Abbé Nollet, remembered for his feud with Benjamin Franklin over the wisdom and efficacy of lightning rods, conducted a number of experiments on people with electricity, with the surprising conclusion: "No inconvenience whatever was felt by persons who submitted to be electrified in this manner. They only found themselves a little exhausted, and had got a better appetite."

Benjamin Franklin applied electric shocks to people who came to him for help, mainly to alleviate paralysis. He gave fairly powerful shocks from a battery of two six-gallon Leyden jars. His patients reported a sensation of warmth, pricking sensations, and limbs that appeared to regain their strength. But Franklin concluded, "I never knew any advantage from electricity in palsies that was permanent." He later realized that the shocks he administered might have been too severe.

Actual lightning injuries were first reported in 1777 by Georg Christoph Lichtenberg, the physics professor who helped introduce Franklin's lightning rods into Germany and who conducted a number of experiments with electricity. Lichtenberg noted that many people who were electro-cuted displayed peculiar, fernlike branching patterns on their skin—the so-called Lichtenberg figures—which ap-peared within hours after the strike and usually disappeared within a day. The patterns, according to Lichtenberg, were the result of an electron shower and not a true burn and, as such, required no special treatment.

In 1794, James Parkinson, a British physician famous for describing the disease that now bears his name, wrote an im-portant article on lightning injuries, in which he described an unusual form of muscle paralysis, later termed ker-aunoparalysis. These lightning victims suffered a total loss of sense and motion in their lower extremities. Nearly a hundred years later, the celebrated French neurologist Jean-Martin Charcot, who had a strong influence on Freud, re-ported in a case study of hysteria "the delirium of the man struck by lightning," who exhibited pallor, trembling, dis-orientation of thought, confusion of speech, repetitiveness, and emotional lability.

Several decades later, Dr. Macdonald Critchley, a London neurologist, described the neurological effects of lightning and electric shock in a 1934 article in *The Lancet*: "As soon as the patient becomes aware of his surroundings [after being hit by lightning] a loss of power as well as of sensation is noticed in the legs and lower half of the trunk. Walking and standing are impossible.... Within the next hour or so—and even less—feeling gradually returns to the limbs, and the patient notices tingling of pins and needles. After about 12 hours sensation and power have usually re-turned and little or no disability can be detected."

"Most victims of lightning either die immediately

or recover . . . without demonstrable long-term damage," Orthello R. Langworthy, a Johns Hopkins University physician, wrote in a medical journal in 1936. As recently as 2001, this opinion was echoed by another Johns Hopkins doctor and two German physicians, who concluded in "The Long Term Consequences of Lightning Injuries," published in the journal *Burns*, that "none of the patients" they evaluated "suffered from any deficits or long-term problems that could be related to the original lightning injury."

For certain, the treatment of patients who have survived a lightning shock is complicated. The diagnosis is easily confused with other causes, especially when a victim is hit while alone in a field or when a disarray of clothes raises suspicions of assault, drunkenness, or some other physical or psychological cause. Even a normal CAT scan, MRI, or EEG result can be misleading. As attorney Simkins warned, "Many physicians and health care providers around the United States seem to operate on the assumption that normalcy on these examinations ends the inquiry as to whether something might be wrong with the person's brain or brain function." Electricity doesn't go to the areas doctors are used to, so they just dismiss it. An electromyogram (EMG), for example, tests mainly motor nerves, not the other peripheral nerves—sensory and autonomic.

If a survivor's EMG is normal, it may mean only that there is no nerve damage to the motor nerves, which send impulses from the brain and spinal cord to all of the muscles in the body. Motor-nerve damage can lead to muscle weakness, difficulty walking or moving the arms, cramps, and spasms. But the sensory nerves could be damaged. They send messages in the other direction—from the muscles back to the spinal cord and the brain. Sensory-nerve damage often results in tingling, numbness, pain, and extreme sensitivity to touch. Or the autonomic nerves might be damaged. They control involuntary or semivoluntary functions, such as heart rate, blood pressure, digestion, and sweating.

"But the EMG is the wrong test," says Cooper. "It's like taking your car to the repair shop, saying I think this *thump thump thump* is the transmission, and it happens to be the flat tire. So the guy looks at the transmission and says the transmission is fine."

Leaning forward in the chair behind her cluttered desk, Cooper goes on to explain that lightning can cause injury primarily in five ways. A direct strike, which is most likely to hit a person in the open who has been unable to find shelter, accounts for only about three to five percent of injuries. A side splash, which occurs when lightning hits a tree, a building, or another person, or when lightning strikes indoor plumbing, telephone wires, or another object and splashes to the victim, is responsible for about a third of injuries. Ground current generated by a lightning strike accounts for about another third; contact injuries sustained when someone touches an object that's been struck account for about one to two percent, and an upward lightning leader, traveling from the ground to the clouds, about a third.

Some injuries may be caused by so-called invisible lightning. During its seventy-five-year lifetime, the Empire State Building has been struck by lightning over a thousand times and has been the object of numerous studies. Over a period of about ten years, several hundred lightning strikes were recorded electrically and photographed, but only about half the strikes produced a visible photographic image. The sub-

The Empire State Building has been struck by lightning more than a thousand times.
(Photo by Greg Geffner)

visible strikes generally had peak currents below 1,000 amperes, compared to brightly luminous strikes that had currents well above 10,000 amps. These results suggest that currents in the range of 100 amperes would likely be subvisible, defined simply as not being visually perceived in daylight by the casual observer.

Neurologists Michael Cherington and Philip Yarnell and biomedical engineer Howard Wachtel of the Lightning Data Center in Denver have investigated this mysterious phenomenon and have suggested that lightning currents across the human chest of as little as 100 amps in strength, lasting mere microseconds, can nonetheless be lethal. Lightning bolts produce intense magnetic fields, which may induce a short-lived strong current in the human body. If this current occurs during a vulnerable part of the cardiac cycle, they have argued, it could cause a severely abnormal heart rhythm and lead to cardiac arrest. Cherington and his coresearchers have suggested that their theory could account for the many unexplained "heart attacks" suffered by hikers, mountain climbers, and campers, even though the usual indicators of a lightning strike are entirely absent. If this theory is correct, many more victims may succumb to subvisible or invisible lightning than to its more visible forms.

Over the years, Cooper has worked to dispel a number of misconceptions about lightning that may prevent victims from receiving the treatment they need. "You may also find that lightning-strike victims suffer from thermal injuries as well as gruesome 'blow out' injuries in direct strikes," states a recent article in *Wildland Firefighter* magazine. "These types of injuries can resemble a gunshot wound: small entry wound and large exit wound." Even survivors search for such signs: "The lightning went in my chest and out my right leg," reports one victim. "The lightning entered through my ear and went out my right toe," says another.

Cooper dismisses such concepts. Pointing out lightning's flashover effect, she argues, "Burns seldom have anything to do with where the lightning entered and exited, because it doesn't enter and exit." By emphasizing entry and exit wounds, she believes, doctors may overlook serious internal or other injuries. Reflecting a disagreement within the medical profession, Dr. Christopher Andrews, an Australian who is another leading medical authority on lightning, insists that he has seen a number of entry and exit wounds on his patients. Dr. Cooper says that she has seen a number of entry and exit wounds as well but exaggerates "on the other side." Otherwise, she adds, people are discounted by physicians and attorneys as having any injury because they "did not have an entry or exit."

Another falsehood is that you can be hit only by a direct strike of lightning. "Every time you read a newspaper article," Cooper says, "the reporter describes a direct strike on a golf course and that the little metal button on the guy's baseball cap attracted the lightning. We estimate that probably only three to five percent of the injuries are direct strikes." In fact, the majority of lightning strikes are caused by side flashes or ground current. The upshot is that by being aware of dangers from such charges, people may take additional safety precautions.

A popular belief is that, after being struck, the victim is still "electrified" and is dangerous to touch. This misconception has led to unnecessary deaths because of delayed resuscitation efforts. But perhaps the most harmful notion, shared by physicians as well as the public, is that if you're not killed by lightning you must be okay, and if there are no outward signs, any injury must not be serious.

A lightning strike may also have some surprising, more-welcome consequences. Some people have claimed that a bolt restored their vision and hearing. Others say that after being struck, they could tolerate extreme temperatures. "I don't notice the cold like most people," observes Harold Deal, a former assistant fire chief in Lawson, Missouri, who was struck in the driveway of his house in 1969. "Some of the folks in town call me Weird Harold. They call me this because I never wear a coat or a long-sleeve shirt no matter what the weather is outside.... I have worked outside in twenty-six degrees below zero, plus I have worked in weather with a wind chill as low as fifty-six degrees below zero. I have also worked in twenty-three degrees below zero for up to seven hours at a time.... A T-shirt and overalls is what I always wear, no matter what the weather is, whether it is summer or winter."

Betty Galvano thinks she is even luckier. Still stunning at the age of seventy-one, with shoulder-length reddish-brown hair and a svelte figure, Galvano modeled in Paris and London in the 1950s and was twenty-three when she married celebrity golfer Phil Galvano in 1958. He hosted the first golf show on TV and later became an instructor and friend of entertainers and sports stars like Frankie Laine, Tony Bennett, and Joe DiMaggio. We are sitting in Galvano's home in Sebring, Florida. From the outside it's like thousands of houses in other subdivisions along the Gulf Coast—single level, stucco, a few palm trees dotting the lawns. Inside, it's crowded with furniture and bric-a-brac.

"A year before I was struck by lightning," she remembers, "I had fallen off a seawall. I lay on the sand for two hours before my daughter heard me screaming. I couldn't walk. The doctors had to put two steel bars in my leg. They are permanently attached to my femur and hipbone. But I still couldn't jump or raise my leg. I couldn't stand on one leg. I couldn't walk without dipping. My leg always felt like a sandbag.

"Then I was hit by lightning one afternoon in June 1994. I was standing there at the kitchen sink cutting up vegetables; my husband was sitting about four feet away from me. It wasn't raining and I didn't hear any thunder. The sun was shining. All of a sudden there was the most unbelievable tremendous blast of thunder. The house shook; the lightning came in through the window. The knife flew one way, the broccoli another, and I slumped over the counter.

"My husband helped me to a couch, where I lay down. I felt like a thousand needles had entered the toes and foot of my right leg." And then an amazing thing happened, she says. "My leg had been sluggish after I broke it, but now it was suddenly full of vitality. I stood up and everything was fine. I could walk normally. When I went back to church on Sunday, the day after the lightning strike, the priest said I had been 'zapped by the Spirit.'

"After church I went to see my doctor. I had no complaints. No memory loss. Nothing. The doctor said I received just what I needed. If I got more, I would have been dead; if I got less, nothing would have happened. There was no damage. I was miraculously healed."

Stories of the unexplained effects of lightning on the body have worked their way into popular culture, perhaps nowhere more graphically than in the bolt-shaped scar on the forehead of the boy wizard Harry Potter. When he was a baby, Harry survived an attack by the evil Voldemort and, as a result, retained the lightning scar. On numerous occasions, he endures searing, blinding pain, "as though his scar were on fire." Harry realizes, though, that his personal lightning bolt provides "a warning...it means danger's coming" or is nearby.

In the *Captain Marvel* comics of the 1940s, news reporter Billy Batson is instantly transformed by a lightning bolt into Captain Marvel whenever he utters the word *SHAZAM!* Captain Marvel is able to harness the power of lightning to fight evildoers. A decade later, Lightning Lad (later Lightning Man), wearing a costume emblazoned with lightning bolts on his chest, possesses the superhuman ability to generate electricity in the form of lightning bolts, which he can project at will through his fingertips. And in 1977, the streetwise character known as Black Lightning, an African-American superhero featured in his own comic book, manipulates lightning to fight crime, injustice, and corruption in the inner city.

Lightning also plays a pervasive role in the story of the X-Men, created in 1963 by Stan Lee and Jack Kirby, the legendary originators of the D.C. Comic superheroes, who went on to create profitable Hollywood films, TV shows, and video and computer games. In the original comic books, a group of teenagers who derive exceptional powers from their mutant X-factor genes are brought together by Charles Xavier, a paraplegic telepathic professor, at his School for Gifted Youngsters on a large country estate in Salem Center, a small town in Westchester County, N.Y. While they are regarded by scientists as the next step in human evolution and by others as a threat to ordinary humans, the X-Men combat everything from sociopathic criminals to galactic forces.

In the "all new, all-different *X-Men*" series created in 1975, the mutants are led by Cyclops (named after the one-eyed giants who gave lightning to Zeus) and Storm, probably most familiar as the character played by Halle Berry in the *X-Men* movies. Gifted with the ability to manipulate the weather, Storm can cause any form of precipitation, generate winds, change the temperature, and direct lightning from her hands, a force she uses successfully. She can also create atmospheric phenomena over a very tiny area, even creating a rainstorm small enough to sprinkle a potted plant. In one comic-book episode, Storm's attraction to dangerous weather leads her to an encounter with Thor, the Norse god of lightning and thunder.

While such characters are clearly fantastic, a number of lightning survivors have claimed electrical or magnetic powers of their own. Gabrielle Blanstein, who was in her home in Haydenville, Massachusetts, when lightning came through an open window in 1996, is supersensitive to electrical current and can feel vibrations when an electrical device is plugged in nearby. So far, she has "demagnetized" two bank cards and a credit card.

Nina Lazzeroni, a motorcycle-safety instructor, was hit in 1995 in Troy, Ohio, after lightning struck a nearby chain-link fence. Her experience left her with an abnormal heart rhythm, burns, numbness, a ruptured eardrum, and nerve damage, among other injuries. But it also left her with the ability to turn off lights at random—streetlights; billboards; lights in buildings, public restrooms, and parking lots. "They come back on after I leave the area and turn off again if I return," she says. "It's not uncommon to have three or four go out while I'm driving somewhere at night. I feel no electricity or any other sensation—I just find myself in the dark periodically."

And Kurt Oppelt, who was an Olympic gold-medal figure-skating winner in 1956, was standing in his bathroom when he was struck in 1989. Seeing his reflection in the mirror, he says, "I looked like Felix the Cat. My hair stood straight up." The next day, doctors could find nothing wrong, but immediately after the strike, "audio receivers, cash registers, and computers broke when I touched them," he claims. "A friend of mine gave me a laptop, but I can't use it. I told my cardiologist that my 'electrical system' was disturbed, but the doctors say they have no scientific proof. I told another doctor what happens when I touch electrical gadgets, and he said, 'So don't touch them.'" Overall, medical experts discount such survivor stories.

Despite the work of Drs. Cooper, Cherington, Yarnell, Andrews, and others, a great deal remains to be learned about how to treat lightning victims and how to mitigate the lasting effects of a strike. Since about two-thirds of lightning-associated deaths occur within an hour of the injury, victims must receive prompt medical care. And since the most common cause of death for lightning victims is cardiac arrest, resuscitation attempts need to be initiated immediately. Patients who are confused or suffer a loss of consciousness should be hospitalized, and a complete physical and neurological examination should be performed, including CAT scans to check for cerebral bleeding; X-rays for broken bones; blood panels to monitor cardiac enzymes; ECGs for the late onset of arrhythmias and changes that might indicate tissue damage in the heart or lungs. (But the majority of survivors need few, if any, of these tests.)

Unfortunately, little is known about which conditions will improve soon after the strike and which will progress to more-serious disabilities. But fortunately, according to Cherington, Yarnell, and Wachtel of the Lightning Data Center, which has produced significant case reports of neurological and other injuries, lightning victims may be helped in the future by innovations like neural prostheses for paralyzed limbs and transplantation of nerve tissue for spinal-cord and brain injuries. Additionally, the increasing availability of portable automatic external defibrillators may improve the chances of recovery for cardiac victims.

For many survivors, it is the unrelenting pain that haunts them. Jerry LeDoux was struck at the industrial plant where he worked in Sulphur, Louisiana, in August 1999. Since then he has pain down his left arm, side, and leg, and severe headaches on the left side of his head. He dozes off to sleep at night, but his body jerks him upright and wakes him, a cycle that repeats itself all night long. Antoinette (Toni) Palmisano has suffered terribly since she was struck by lightning in her office in an insurance agency in Mt. Pleasant, South Carolina, in May 1991. "My lower-back pain is so disabling," she says. "I cannot sleep on my back or on my stomach. My arms have such deep pain all the time, I can't even hold a pencil to write my name. Daily, I can barely open my eyes due to the headaches and pains in my eyes. My ears have a high ringing sound constantly. My legs and parts of my body experience daily 'hot spots' and weakness, causing me to fall frequently." She complains, too, of the feeling that there are "bugs crawling under my skin" along the left side of her head, her shoulder, arm, hand, and down the side of her body to her toes. Some victims find temporary relief in morphine and other drugs; some never do.

Of all the research he would like to see undertaken, Steve Marshburn, Sr., the founder of Lightning Strike and Electric Shock Survivors International, first mentions a remedy for the pain that "no one knows how to explain. It's a pain that makes you feel like your head will blow out because of the headaches," he says, "like your back will blow out, your legs, your feet. Sometimes it's in your extremities. Your eyes hurt so badly. This pain is one that goes from bottom to top." He would like doctors and researchers to determine why the brain is damaged and why the vertebrae are affected. "Why are there crushed discs, blown discs, bulging discs? It's all because of lightning surging through our bodies, but why?"

At this point in the medical literature, there is not even a suggestion of a cure for brain damage suffered from lightning. The principal treatment is cognitive therapy, a kind of retraining of the brain's thought processes. After two or three years, many survivors reach a point where they accept the injury and its challenges but don't make it the focal point of their lives. "They know it's there. They know it will always be there. But they don't let it control their lives anymore. They control it instead." Cooper emphasizes the opportunity that lightning survivors have to change the direction of their lives.

"Your grandmother had a teacup that was your favorite. One day the cup gets knocked off the shelf and breaks, but you can't bear to throw it away. So you get all the pieces and put it back together again. That teacup is never going to be strong enough to hold tea again, but it can still hold your keys or a plant. It can still have a use.

"That's what happens to people who have a lightning injury. They are always going to see the cracks, but they can live a happy and productive life.

"What we need to do," she suggests, "is to find something that would stem the cascade of injury. Is there something that can be done in the ambulance? During the first day? During the first two weeks? Is there something that could halt and reverse nerve injury?"

Asked about her priorities for future research, Cooper muses, "If I were twenty years younger, I would like to be able to find something that would mitigate the brain injury and nervous-system injury we see from lightning. Yes, I'd like to take off in that direction. I'd learn a lot and maybe make a difference."

A GATHERING OF ANGELS

*The fire of God is fallen from
heaven and hath burned up the
sheep, and the servants, and
consumed them; and I only am
escaped alone to tell thee.*
—JOB 1:16

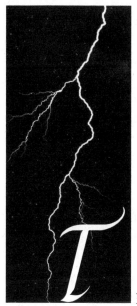hroughout Dollywood, a kind of
rustic Disneyland nestled at the foot of the Great Smoky
Mountains in Pigeon Forge, Tennessee, Dolly Parton's trademark bosom and equally inflated blond hairdo appear on
posters everywhere, hovering over visitors eating at Pop's
Catfish and Seafood, touring the Elvis Museum, viewing the
moonshine stills at the Hill-Billy Village, enjoying gospel and
patriotic music at the Country Tonite Theatre, tying the knot
at the Wedding Bell Chapel, or settling in at the Miracle
Theater to watch a "magnificent battle of angels" and "the
birth of Christ in all His glory." The most popular attraction in Pigeon Forge, by far, however, is Dolly's own Dixie
Stampede, where the audience chows down on hickory-
smoked barbecued pork loin and hot homemade buttermilk
biscuits while watching performers outfitted in Union and
Confederate uniforms vie in rodeo-style reenactments of

skirmishes from the Civil War. At Christmastime, it's the North Pole vs. the South Pole.

Meanwhile, at the nearby MainStay Suites Hotel, the annual world conference of Lightning Strike and Electric Shock Survivors International is taking place, and men and women who know the devastating effects of lightning first-hand are pouring out their hearts and souls in hopes of restoring their broken spirits and shattered dreams. Inside the hotel, Ron "Griz" Swidorsky is careful not to squeeze too hard when he shakes a stranger's hand. He doesn't know his own strength, he says. His large strong hands don't look as though they could strum a guitar or manipulate a drummer's sticks or play five different instruments, but for most of his fifty-four years, Griz has crisscrossed the country playing and singing everything from Western music to blues and oldies. "There are times when music pulls me down," he admits sadly, "and I stay there for a few days, because it takes me back to what happened—to the lightning."

Griz was felled by lightning fourteen years ago when he grabbed the door handle of his Ford leisure van to roll up the windows as a storm was approaching. An extroverted, bearlike man whose nickname, "Griz," recalls the time when he used to hunt bears in northern Ontario, he remembers that after the strike he felt like he was "on fire" and he couldn't use one arm. (Actually, he had been hit once before, in 1977, while he was unloading a truck, but he shook it off at the time, saying, "What the hell was that?") Since the second strike, he's had numerous health problems and a lot of suffering—his nerves are damaged, his left side hurts, and he gets depressed. But he has worked through the pain. "I used to feel sorry for myself, but I don't anymore," he says. "I can take things most people can't take. My buddy had his finger cut off and I took him to the hospital. I carried the finger in a bag. I can handle any situation. I was always like that, but I'm even stronger today."

The Swiss psychiatrist Elisabeth Kübler-Ross has identified five stages of grief that a dying patient experiences, and lightning expert Dr. Mary Ann Cooper believes that these psychological steps apply to lightning survivors as well but notes that they may not be experienced in a particular order. First, there's denial ("This isn't happening to me!"), then anger ("Why is this happening to me?"), often displaced onto physicians and families because the victims can rarely find anyone else to blame. Next comes bargaining ("I promise I'll be a better person if I recover," or "If I just find the right drug, the right exercise, the right therapy, I'll be back the way I used to be"). After that there's depression ("This isn't ever going to go away. I just don't care anymore"). With a broken leg or some other common complaint, there's a predictable length of time for recovery, but with lightning injuries it may be months, or years—or never—before these victims get better. Depression sinks in. At the recent meeting of survivors, one man threatened to kill himself, but by the end of the three-day conference, other survivors, family members of survivors, and a therapist had helped him work through his suicidal thoughts. Finally, if the victims are lucky, they may achieve the fifth stage, acceptance ("This isn't going to control my life anymore. I can move on. I'm ready for whatever comes. I can handle this"). Like Griz.

"The day my life changed forever" is a common theme that weaves through the experiences of lightning survivors.

They suffer pain, crippled limbs, loss of memory. They are fired from their jobs, their marriages break up, their friends disappear, their dreams vanish. They suffer from depression, guilt, anguish, anxiety, and fear. Their physical, psychological, and spiritual lives change greatly. But while survivors may not be able to return to the way things were before their injury, they can also experiment and create new possibilities for themselves and strengthen their families.

Dr. Cooper recalls a patient who was abandoned by his friends after his lightning injuries and had trouble working. But he was still able to go out and walk his dogs every day. The dogs didn't care that he had a brain injury. They didn't care that he moved more slowly. Eventually, the man changed his career and became a professional dog trainer.

Though the eminent psychiatrist Robert Jay Lifton and others have written about the psychological trauma caused by natural disasters, there is no body of research on the similar effects of lightning strikes. What data there is focuses mainly on the medical and neurological rather than on the emotional consequences. In *A Match to the Heart*, Gretel Ehrlich describes the ways in which her life changed after she was hit by lightning. After attending a meeting more than ten years ago of the lightning strike survivors' organization, she wrote, "I thought of those humans who had awakened after being hit and became shamans and healers, and wondered what this new life of mine would be, carved from a ruined body and a ruined marriage, and what special passageways I could hollow out as in a labyrinth of dead ends." But her book is an unusual example of one survivor's story, rich in detail. The other outstanding examples are the three volumes of *Life After Shock*, published since 1996 by Lightning Strike and Electric Shock

Survivors International, Inc., which are personal stories of survivors.

An avid scuba diver, hiker, and mountain climber, Judith Friend, a pharmacist, was struck by lightning in 1979 while backpacking with her husband for her twenty-seventh birthday through the Rawaw Wilderness near Fort Collins, Colorado. As her ailments lingered year after year, she began to feel as if her body had given up on her. But she slowly pursued her recovery. "I was too young to feel so old," she says. She and her husband have now trekked through Thailand, bicycled in China, and backpacked in Peru. "In a weird way these things can be positive," she observes. "You might not be able to climb Mount Everest, but there are other facets of your life you can enjoy. . . . Spiritually and emotionally you come out stronger. There is life after lightning."

Immediately after a strike, particularly when there is serious injury like heart failure, loss of consciousness, or shock, the survivor wonders: "What happened? What did I experience?" Of course, such confusion can overcome the victim of any trauma or sudden tragedy as well.

A survivor also tries to understand why he or she has been singled out. Lightning is a unique and rare kind of accident in which the victim in most cases bears absolutely no responsibility for what has happened. Those few individuals who are struck every year cannot overlook the fact that they have received what seems to be, literally, a sign from the heavens. The question passes briefly through the mind of even the most secular, nonreligious victim: "Why did this happen to me?"

Many survivors, grateful for being spared from death, believe they have an unfulfilled mission in life. No matter how he or she has lived in the past, the question becomes: "What is expected of me now?"

*S*ome survivors report having a near-death or out-of-body experience right after they were struck, similar to experiences unrelated to lightning recounted in the groundbreaking book *Life After Life*, by doctor and philosopher Raymond Moody. During such an episode, "one seems to be no longer in one's own body but outside it," neurologist Dr. Oliver Sacks observes more recently, "and, most commonly, looking down on oneself from eight or nine feet above." People who undergo such experiences have the sensation of "floating" or "flying." They recall traveling through a dark tunnel; seeing a bluish-white light; conversing with deceased relatives; feeling joy, ecstasy, and extreme peace; and, above all, experiencing a sense that something deeply significant and transcendent has happened.

Out-of-body experiences are mystical experiences, taking place seemingly beyond space and time. Though they may be difficult for others to comprehend, Dr. Sacks points out, such events "are not easily dismissed by those who have been through them." Typically, they may "lead to a conversion or metanoia, a change of mind, that alters the direction and orientation of a life." Such experiences, again, are not limited to lightning survivors but have been shared by others in situations of great danger or trauma.

*E*rin Ellickson was an eighteen-year-old sophomore at the University of Florida at Pensacola when she and her parents were hit by lightning in August 2004 as they were walking down the steps from her dormitory. Erin was holding a cell phone at the time. All three were knocked unconscious by the strike, but her parents quickly revived and realized that their daughter was not breathing and that her heart had stopped. "We crawled over to her," her father Ed says, "and tried to do CPR, but I couldn't remember what the steps were. A young man who was walking nearby and was trained in CPR took over and brought Erin back to life," he recalls gratefully. "They carried her inside, but again her heart stopped. But the man kept giving her CPR, and eventually the police arrived and the EMTs and they got her to the hospital."

Erin suffered brain damage, similar to a stroke, and spent eight weeks in the hospital. She has had to relearn how to walk, talk, eat, and even breathe, and she is sometimes hard to understand. "She's had some weird behavior," says her father. "You never know what's going to happen next. She's violent, sweet, loving, and at times does outrageous things like walking out the door with nothing on. When she got injured the brain disconnected, and now it's trying to connect again. Like rebooting your computer."

Immediately after she was struck, Erin had a near-death experience. "I saw God," she says. "He was talking to me. He said I was struck by lightning. I was in awe. I said to Him, 'Get off your ass and start doing something.'"

Then she saw her grandmother, who had died the year before. Her grandmother said, "'It isn't time for you to be up here. Get back down where you're supposed to be.' It was a nice experience," Erin recalls, "but I was confused."

Two schools of thought have tried to explain the

near-death phenomenon. One, characterized as the "after-life hypothesis," suggests that a near-death experience is in fact an actual glimpse of an existence after death, a mind-boggling but accurate account of what awaits us all at the end of life. The other, the "dying-brain hypothesis," suggests that the visions and sensations that accompany a near-death experience are purely physical phenomena, symptoms of terminal mental states that will ultimately stop when the brain ceases to function and that tell us nothing at all about a possible life after death. Dr. Sacks, whose lifework has involved the study of the mind and brain, concludes, "We have, as yet, little idea of the actual neural correlates of such experiences, but the alterations of consciousness and emotion that occur are very profound and must involve the emotional parts of the brain, including the amygdala and brain-stem nuclei—as well as the cortex." The important point, as leading researcher Susan Blackmore writes, is that a near-death experience "can change a person's life forever but it is not necessarily evidence for life after death."

*I*t sounded like a bomb hitting the ground," remembers Lindsey Sass-Aurand of the lightning bolt that struck her and her husband, Paul, in the summer of 1998. They had been sitting on the beach at Greenwood Lake in upstate New Jersey, watching their son's swimming meet. When a thunderstorm began moving rapidly across the lake on that July afternoon, her other son, three-year-old Orion, started to cry. "It's all right," Lindsey remembers telling him. "We can stay. I'm not afraid of thunderstorms." But as the toddler continued to wail, the couple walked up the hill toward their car. As Paul was putting Orion inside, Lindsey felt the hair on her arms stand up. That's when it hit.

"Everything went black, and I have a hard time saying this, but I saw a light and my father was there and he was guiding me," Lindsey says, although her father had been dead for many years. "I was moving through layers of energy toward the light but I didn't make it, as I was stopped by two very large shadows who showed me lost spirits that seemed to be floating around me. Maybe they were people who had died with no belief system and they weren't ready for the light.

"I felt like a tiny piece of sand," she says quietly, "very small compared to this very large shadow. Some people would call it an angel. When it embraced me with its energy, I never felt more loved. It was an unconditional love.

"I wanted to stay there, but the shadows said it would be too hard on my family. It's not that they forced me to go back. They gave me a choice. Then I was going through a tunnel. I was traveling very fast. It was like a vortex, pulling me down. I remember being back in my body and I was in a contorted, paralyzed position on the ground. Orion was in the car and he was crying, and Paul was screaming, screaming at the top of his lungs."

Regaining consciousness, Lindsey saw "angels helping us" in the mist around her. "Everyone else was yelling, 'Call 911, call 911,' and I'm yelling, 'Call the angels, call the angels,' because I could see them," she said.

During her near-death experience, Lindsey felt that she had gotten "stuck." "I was the kind of person who was unforgiving and resentful," she admits, and after her injury, she felt she had to change and become a more generous person.

Her husband, Paul, a holistic healer and hypnotherapist, who was also struck, later recounted an episode that occurred well after the strike while he was participating in a therapy session. "Everything grew dark," he writes, "and I entered that great void, that 'nothing' place people report in their near-death experiences." When he opened his eyes, he felt "completely transformed ... free of the chronic back pain I had experienced and far less fearful of electric storms." He adds that among the questions he "lives with every day" are: "Can I ever be the same again? Do I want to be the same again?"

Griz had his own near-death experience after he was walloped while holding on to the door handle of his van. "I'm not a believer in the spiritual stuff," he tells me, "and I'm not a believer in the afterlife. But that night in the hospital, I kept seeing people that were gone. My grandma, for one. She was pointing her finger at me the way she did when I was young. I also saw a grade-school friend of mine, Daniel Pewter, who had gone into an abandoned house a few days after a fire and was killed when part of the house collapsed. We were fourteen or fifteen at the time. I also saw this thing on the wall, like a script. On it were all these things that I did wrong in my life and things that I did good in my life and things I wanted to do in the future."

Since then, he has accomplished everything on the "to-do list" that was revealed to him during his near-death experience. He wanted to earn a high-school diploma—he had dropped out of school to hit the road as a musician—and he did. He wanted to obtain a motorcycle license—and he did. He wanted to design tattoos that commemorated his life—

and he did. The fifteen tattoos that decorate his body now include a bear, a rainbow trout, his dogs, and a lightning bolt with the date when he was struck. (I've discovered that many survivors get tattoos of lightning.)

Like other survivors, Griz wrestles with the question "Why me? Why did God do this to me? I've been a good guy. I've done a few bad things as a kid. We've all done that. I've never hurt anybody. I've never killed anybody. There's people I know who are bad people. They seem to go on and nothing bad ever seems to happen to them, but they do bad things constantly. It's not fair. I feel like I'm being punished for something I never did.

"I've become a better person," Griz says about himself after the strike. "I'm always there to help people. From what I've been told, I've inspired a lot of people to get on with things in their lives that they never thought they could do."

*I*n their search for answers, victims often experience a greater commitment to faith, religion, and a heightened sense of spirituality, or they simply become stronger, kinder, more patient and tolerant. When I talked with Betty Galvano at her home in Sebring, Florida, about whether being struck by lightning had changed her in any way, she told me, "I realized that the only thing that matters in life is loving God and loving each other. I've always loved God. But since I was hit I feel totally immersed in God. He is with me all the time.

"I don't fear death. I live with one foot in this world and one foot in the other world. Just knowing that God's here all the time, there's no reason to worry. I don't worry. I pray.

I've always had this feeling that angels are nearby. They're all around."

Galvano seems to have had an uncommon relationship with storms since she was born in Atlanta, Georgia, in 1935. "When my mother brought me home from the hospital, there was a tremendous thunderstorm. An electrical cable had broken outside, and the electricity started jumping from the window to doorknobs. The whole house was filled with electricity. They wrapped me in a raincoat and took me out of the house.

"I was a blue baby. My head was like an eggshell and my circulation was poor. When I was three months old, I was in a coma in a hospital where a doctor told my mother that there was nothing more he could do and that I would not live beyond the next three hours." Meanwhile, a terrible thunderstorm was raging. "My mother looked out the window, and on the lawn she saw a ball of lightning. It floated up, came through the window, and rested on top of me. At that moment a nurse came into the room and watched with her mouth wide open as I came out of the coma. The nurse ran out of the room screaming, 'Miracle! Miracle!'

"When I was three years old," she continues her amazing saga, "I was in a coma again. My parents and a minister were standing around me and the doctor said, 'By five o'clock she will be dead.' Again there was a lightning storm outside. All of a sudden the lightning stopped, and I came out of the coma and looked out the window and saw a beautiful lady and children dancing around a tree. I know I saw an angel when I saw this lady with wings dancing around the tree. I've always had this feeling that angels are nearby. You should talk to them. They are all around you."

When I tell her that I'm Jewish, she says, "Jews have angels too."

"You've had a special relationship to lightning," I respond. "Tell me about the time you were struck."

After a bolt crashed into her through an open window in 1994 while she was standing in her kitchen in Fort Myers, Florida, cutting fresh broccoli, she began making dolls that she called "Joyful Angels." She made many different kinds, including Native American angels and golf angels. She buys the ceramic heads and glues them onto cardboard that's cut into a cone shape. Each angel is about ten inches high. "After four months I had them in forty-two states. I've sold hundreds for about sixty dollars apiece. It was a very lucrative thing. I sent one to Pope John Paul, who sent me an apostolic blessing. But I stopped when my husband died in 1996." After the strike, Galvano also built a spiritual garden at the St. Francis Xavier Church in Fort Myers. "Everything that's planted there is mentioned in the Bible," she assures me. "There are statues, stations of the cross—twenty-six angels in that garden."

Galvano's house is full of religious objects. An angel adorns a red candle in the center of the black dining-room table. Seven angels hang from a chandelier above. Angels are everywhere: ceramic angels in the bathroom, painted angels on a dome in the foyer, an angel statue in the garden. A hundred angels are scattered throughout the house, not including the ones Galvano made herself.

In her foyer can be found a wooden pew; a two-hundred-year-old statue of the crucifixion; a ceramic St. Joseph ("I found that in New Jersey"); a relic of St. Jude; a picture of Jesus and an angel comforting each other ("My favorite picture, I found it at a garage sale for twenty-five cents"); a four-foot-high statue of St. Teresa; a statue of Gabriel; a St. Joseph's staff made of wood; a picture of the Infant Jesus of Prague.

"I must have about fifty statues," she says. "I feel they are all my brothers and sisters."

As I say good-bye, she points to the doorpost. "I even have a mezuzah."

*J*effrey Bryer lives in West Boca Raton, Florida, "where the Everglades meet civilization," he says. "We get a lot of lightning out here."

We are sitting at a table in the shade at an equestrian center in Delray Beach, Florida. The sun is shining brightly this beautiful January morning, the temperature in the high seventies. Bryer is of medium height, with brown curly hair and a goatee, and he's dressed casually in a polo shirt and shorts. He waves to his seven-year-old daughter, who is on horseback. In the distance, sulky drivers are exercising their trotting horses. In colorful shirts and hats, they move in fixed circles around the track.

Bryer, a professional entertainer who sings, plays the piano, and fronts an orchestra, has been beset by serious accidents throughout his life. When he was four, he was nearly killed when he fell down a double stairwell on his tricycle. When he was thirteen, his bicycle was struck by a car. He caromed fifty feet through the air, was knocked unconscious, and spent seven days in the hospital.

"When I was twenty-one," he says, "the engine of a truck I was driving exploded in the middle of a highway . . .

gasoline, flames...the vehicle rolled to the median. Lots of blood, lots of injuries. I was impaled on a window crank and couldn't get out, and I said, 'Please, God, let me out,' and the windshield popped like a champagne cork, and I crawled through it. A few years later I was skiing, just north of Toronto. I was going down a mountain and lost control and splattered into a six-by-six-foot fence."

Riders canter past. A radio is playing country music in one of the stalls. Bryer cracks his knuckles. A relative—a great-great-great-grandfather or a great-great-uncle—fought for the Confederacy and was one of the first Jewish soldiers killed in the Civil War, he believes.

"In 2001, I was driving on the turnpike, and lightning was striking everywhere—trees, cars—like bombs falling from the sky. I was frightened. Lightning hit the black car in front of me. The car kept going, but it turned bright red, bright red. I'll never forget that."

Then, in June 2005, as he was standing in his garage in bare feet on a cement floor holding a vacuum-cleaner hose, lightning struck him. There were only a few clouds in the sky at the time, no sign of a thunderstorm.

"Lightning hit the front corner of my neighbor's house," he says, "struck a palm tree in his yard, and then came across the street into my garage. It traveled horizontally and came right into me. The jolt sent me up to the ceiling—about twelve feet. I came back down on the ground. My hands were burned. My heart was racing. I couldn't hear anything. The lightning had set off all the car alarms on the street. I couldn't hear for a couple of hours, and I was in a lot of pain. My wrists and forearms were numb. A neurological specialist said my nerves were all inflamed.

"As far as waking up and seeing the light or being a holy roller? That didn't happen. But lightning may have made me more appreciative of being alive. I suppose I take an extra breath and count ten seconds before I get really angry at someone. When I react in a negative way, karma, it comes right back at me. If somebody cuts me off on the road, I say a prayer for them.

"I don't know why I'm here. There must be a reason. . . . I think my kids are probably one of them. I say a lot of prayers for my kids, my family, and for the world. All these experiences—are they a wake-up call? Or is it just something random? Many times I thought to myself: Why am I still here? My answer is, God has a reason. He has a plan for me."

*O*n a July afternoon in 2006, Veronica "Ronnie" Smith, a retired nurse's assistant, was volunteering at a Bible-school camp at the Second Baptist Church in Worcester, Massachusetts. Smith was afraid of lightning. She'd been influenced by her mother's stories of growing up in the South, where her mother's parents followed the common custom of always turning the TV and lights off when there was a thunderstorm. The whole family had to sit in a darkened living room with all the shades drawn. "God is doing His work," her mother had said. "You have to be quiet and still."

We are sitting in the comfortable living room of her brother's house, near Hartford, Connecticut. Several paintings on the walls depict black soldiers who fought for the North during the Civil War.

Smith said that although the rain had stopped outside the church on that July afternoon, she was still apprehensive and told the children to go inside. "Okay, we're coming, old superstitious lady," they shouted, and then remained outside. (She now calls them "my little angels.") Lightning hit a nearby tree and swept over her as she was using her cell phone. She was lifeless and blue in the face. A bystander tried CPR but couldn't get her to breathe. Her breathing returned only after paramedics arrived and used a defibrillator. The first thing she remembered after being hit was "floating alone. Then I saw an image that was swaying. It was my sister, an older sister whom I was very close to and who had passed away from cancer. She looked at me and said, 'Ronnie, it's not time. Not yet.'"

Smith didn't wake up until the first night in the hospital, where she spent three days suffering from burns over her body, pain, loss of hearing, and a burning sensation. She was wearing a cross, and an imprint of the necklace that held it was outlined on her chest. She is still in terrible pain.

When she asked her pastor why she had been the victim of lightning, he told her, "It wasn't meant for you. If it was meant for you, you wouldn't be here today."

But she gives her own answer: "There was nothing that I had done that I could be punished for. I believe that God was trying to get a point across, because there had been arguments among members of the church. They weren't together. After my accident they came together. I helped out. It was a way of God telling us that it was time to love one another and to get along."

*T*here is a different lesson for everyone," says Ed Ellickson. "In my case, it's patience. We're not an outgoing family. We don't have a cluster of friends. But since this ordeal we've found it a lot easier to talk about things. We've ended up leaving our name and number at various facilities."

He mentions the 2006 Sago Mine explosion, which killed twelve workers when lightning traveled two miles aboveground before igniting methane gas in an abandoned section of the mine. "I even called the coal miner in West Virginia who was stuck in the mine and basically came out with the same problems as Erin. We told the miner, 'If you ever need to talk to us, give us a call.' Perhaps that's our calling. If that was God's will, I don't know. I think we are a lot more patient, a lot more caring about other people's problems now."

*N*athan Loyet wears the tattoo of a thunderbolt encircling his right arm like a badge of honor. He is sitting slumped in an easy chair in his comfortable home in Collinsville, Illinois, across the Mississippi River from nearby St. Louis. He is twenty-four years old, but he looks and acts much younger.

On a beam across the living-room ceiling, ceramic angels rest. "I have angels everywhere, indoors and outdoors," explains his father. "I believe in angels."

Nathan was only thirteen years old when he was struck by lightning at his home. When he reached to close a garage door that had blown open, a bluish bolt traveled from his hand, up his arm, and across his chest. He remembers "a weird smell, something burning, like the TV when it goes off." His skin turned brown, his veins were bulging, and bumps and welts appeared on his chest. But after a series of tests, he was released from the hospital with the note: "Struck by lightning, may return to school. Take two Tylenol every four hours for pain as needed."

When he entered the eighth grade a few days later, he had numbness in his hands, occasional sharp pains, fatigue, and memory problems. He soon displayed behavioral problems as well: fell asleep in class, was expelled, and briefly ran away from home. His parents, Becky and Paul Loyet, were frantic and didn't know what to do.

During the summer of '96, when he was struck, Nathan had spent most of his time looking for arrowheads near his home on the Mississippi River floodplain. The site was first settled by Indians in about 700 A.D., then abandoned in 1400. But in the late 1600s the Cahokia Indians came to the area, which is defined by some eighty mounds that the earlier Indians had constructed for unknown reasons. Near the Cahokia Mounds, Nathan found arrowheads, spearheads, scrapers, and also a strange three-dimensional pottery face, which he stored in a little cigar box in his closet. Archaeologists identified it as a genuine piece of ancient Indian pottery, believed to depict Tlaloc, the wrathful Aztec deity of lightning. "It was scary," says his mother, Becky. "The first time I saw it I got the chills."

Immediately after Nathan was struck, he had an out-of-body experience, during which he had the sensation that he was on his property walking among Indians and Indian tents but he was invisible. Later, he had recurrent nightmares that he was floating down a creek in a wooden boat and lightning kept striking him. There was a big tree and "Indians were there, and I'd steal their arrowheads and then start floating up to the sky, but I resisted and tried to hang on to the tree."

When he had these dreams, he would scream and hit his headboard with his fists. "He was fighting the lightning," his mother says.

To relieve his nightmares, pain, frequent vomiting, and other problems, Nathan's parents took him to more than twenty doctors and specialists, to a chiropractor, to an alternative-energy healer who provided acupuncture and other treatments. One day a Native American woman contacted Becky Loyet, asking if she was interested in knowing about the Native American approach to lightning. After the Indian woman and Nathan's mother had corresponded for a year, the woman was invited to the Loyet home to cleanse it of evil spirits. In the summer of 2001, Julie Spotted Eagle Horse, a Lakota Sioux healer, arrived with another Native American healer.

Julie Spotted Eagle Horse had a special interest in Nathan Loyet, as she, too, had been struck by lightning— twice. The first time was in the late 1990s when a bolt came through wiring and a dryer vent as she was doing laundry in her mobile home in West Burlington, Iowa. The second time, in the fall of 2000, almost exactly on the date of the first strike, lightning came in through an open window in her Burlington house. Primarily from the first strike, she suffered, among other things, burns, cognitive damage, and memory loss. She told a Lakota holy man that strange things were happening, and he suggested that she pray and take part in a sweat-lodge ceremony. During the ceremony, she saw a Wakinyan (Thunderbird). "It was a black bird, as black as can be," she says. "It was the size of a human being, with energy coming off its wings."

The Sioux believe that the moment an Indian gets

struck or dreams of lightning or thunder or of Wakinyan, he or she becomes Heyoka, somewhat like a medicine man but one who is "contrary," who does everything backward. Heyoka is both sacred and ridiculous, sorrow and laughter rolled into one. People fear Heyoka, and even the person is a little afraid of himself and his strange powers, which derive from the Thunderbirds. After finishing the ceremony, the medicine man said that Julie Spotted Eagle Horse had become Heyoka.

Now she and the other Indians built a sweat lodge in the woods below the Loyets' house, where they conducted sacred ceremonies for Nathan, but he did not participate in them. A few days later, Sister Barbara, a shaman from the Pine Ridge Reservation in South Dakota, arrived with a "light worker," performed rites, and said prayers. At the insistence of the Indians, the Loyets destroyed the strange pottery face, without telling their son.

A month or so after the piece was destroyed, Nathan graduated from high school. He seemed calmer. "It was a form of closure," his mother says. One of the Indians had predicted that Paul Loyet was soon going to save many lives. Shortly after, he went to work for the Red Cross. Becky says that through Nathan's experiences she learned to help other children and she took on additional responsibilities at a local parochial school and in her community. She has also become a key member of the lightning strike survivors' organization.

Except for lingering problems of fatigue, Nathan is now in good health. Before I leave, he cheerfully drives me in his pickup to see Collinsville's best-known landmark, after the Cahokia Mounds: an old red, white, and blue water tower, a hundred and seventy feet high, built in the shape and colors of a giant catsup bottle. Allegedly, it has never been struck by lightning.

*N*ative Americans believe that when a human being is injured by lightning, there is a reason, though it may be unclear. They also believe, as did the ancient Romans, that when a person is killed by lightning, he has done something to challenge fate. I thought of this as I learned Tim Teson's story.

Tim was born and grew up in the Tom Sawyer town of Hazelwood, Missouri, where the two great rivers of North America—the Mississippi and Missouri—become indivisible in their long passage to the Gulf of Mexico. Hazelwood is only a few miles away from St. Charles, which was founded in 1769 and is the second-oldest city west of the Mississippi. Daniel Boone settled in the region, and in 1804 Lewis and Clark set out from St. Charles on their historic expedition.

On his father's side, Tim's ancestors were French trappers who arrived in the area in the 1700s; on his mother's side, farmers who had emigrated from Germany at the end of the nineteenth century.

Tim was born in 1964 and grew up in the '70s and '80s. He spent many happy weekends with his parents, brother, and sister camping on sandbars, where they swam, water-skied, played volleyball, and gathered around campfires on summer evenings. Sometimes they camped on the Missouri, sometimes on the Mississippi. The Mississippi was the more refined, with its sloping sandy shores and gentle currents;

the Missouri was uncouth, with its treacherous currents and muddy banks.

Even after Tim and his brother and sister grew up and left home, the sandbars were the places where the family gathered on weekends. "We loved the river," recalls his mother, Judy, "but it was our youngest, Timmy, who had the most affection for it." Later, when he went away and joined the Coast Guard, he missed it so much that he wrote home: "When I get out, I'm going to live on the river for one month straight."

From the time Tim was a child, however, his father says, "I wondered if there was a black cloud hanging over that boy's head." When Tim was six, he fell and had a concussion. Then one day the family was camping on a sandbar and a terrible thunderstorm flared up. To this day, townsfolk remember it as the "hundred-year storm." Tim's parents, who were Lutherans, huddled under the tent and prayed. "It scared Timmy so bad," his mother remembers. "It took five or six years before he went near the rivers. He never forgot that experience."

As a teenager, Tim returned to the Missouri and the Mississippi and grew into a six-foot-two-inch, broad-shouldered, ruggedly handsome man. At eighteen he fell in love with a woman named Cindy. His parents tried to talk him out of marrying.

"I finally persuaded him to join the Coast Guard," his mother comments. "I thought it would be a way to break up the relationship." But Tim married Cindy during Christmas week of 1983. On New Year's Day, they left Hazelwood for Tim's basic training in Cape May, New Jersey, towing a U-Haul trailer behind them.

Tim served two years on an oceangoing cutter, where he eventually lost his fear of storms. "Mom, you have no idea what storms are," he wrote, "until you see one in the Atlantic Ocean." He weathered many gales at sea, with winds that tilted his ship fifty degrees and waves that reached twenty-five feet. A Lutheran who rarely went to church, Tim nevertheless took a dog-eared Bible with him.

When he had completed his service and returned home, Tim often chided his parents when they were waiting for a storm to pass over. "You bunch of pussies," he would call out while standing chest high in the river, drinking beer in the pouring rain. Tim was intense and passionate, his mother says. "He had to do everything at a hundred-ten percent."

In his first job after the Coast Guard, Tim worked on barges that floated down the Mississippi to New Orleans. After a few months, he returned home and went to work as a production supervisor at a factory making refrigeration equipment. Though Tim's wife had given birth to a baby, their relationship had soured.

Tim eventually filed for divorce and fell in love again, with Karen Henkel, a young woman he met at work. The couple used to go camping often, and during the first week-end of August 1994, Tim, Karen, and his parents spent a day on a sandbar on the Missouri. As his parents were getting ready to leave, Tim was standing at the water's edge when his mother gave him her usual warning: "If a storm comes up, you be careful."

Annoyed at hearing the same old tiresome advice, Tim raised both arms to the sky and, holding a beer can in one hand, shouted, "Strike me, lightning, strike me!"

On August 13, a week after Tim had issued his challenge, he and Karen took their boat out on the Missouri. Usually there were other couples camping out on the sand-bar, which was about the size of two football fields, but on that particular weekend Tim and Karen were alone.

The first storm blew in at about eight p.m. There was a lot of wind but little rain or lightning. The couple laughed and laughed as the wind intensified. "Boy, if Mom were here, she'd be freakin' out," Tim said.

As the winds died down after about an hour, they strolled along the sandbar, but when they were returning to their tent, Tim noticed an ominous lightning show in the distance. Though Tim would go out of his way to avoid ele-vators and airplanes, he had no fear of thunderstorms; still, he asked Karen if she wanted to go back. "I just want to stay here with you," she said.

At about ten p.m., the worst thunderstorm of the year was on top of them. As thunderclaps shook the ground and rain poured down in sheets, Tim and Karen lay side by side inside their tent on an air mattress. The lightning was so in-tense that the sky had an eerie orange glow that could be seen a hundred miles away. As the rain leaked through the tent, Tim pulled on a blanket and joked, "All we need now is a tornado."

At about ten-thirty, a bright flash—brighter than a welder's arc—tore into the tent. Karen screamed, trembling and hysterical. She had an inch-wide bright-purplish streak running from her left knee up to her chest and down each arm, and both hands burned. The shock had paralyzed her right arm and hand, and she tried desperately to slap life back into them. The flash had blinded her right eye, and the explosion had pierced a hole in her right eardrum and left the other ear partially deaf. She later said it was like hearing sounds underwater. She reached over to see if Tim was all right, but he didn't move. She tried the tent's zipper, but lightning had melted it. With her burned but still-useful left hand, she finally ripped a hole in the tent large enough to crawl through.

Karen ran to the boat, a sixteen-foot outboard, and pulled up the anchor. She pushed the boat from the sandbar, jumped in, and turned the ignition key. Except for flashes of lightning, the night was pitch-black, and the rain, still coming down in sheets, was pelting her good eye. She had never steered the boat by herself before. If she wandered out of the channel, she would likely run over a dike and tear off the bottom of the boat. If she hit a buoy in the dark, she could be killed. She had to rely on everything Tim had taught her only that afternoon. As she made her way up the river, she screamed for help, but there was no one around to hear.

When Karen reached the landing, she threw the anchor on the muddy bank and ran to Tim's car. Steering with her left hand, she sped off and drove to the nearest gas station, where she found a phone booth and dialed 911. When the rescuers arrived, they told Karen that the lightning was too intense to launch their boat. Karen pleaded and screamed at them. "He needs help. He's hurt. Please help him."

Reluctantly, they gave in to her pleading and set their boat into the river, risking their own lives. By the time they arrived at the sandbar, the tent had collapsed. The blanket, still covering Tim, bore a long, melted black streak. Tim had a reddish-brown circular burn on the inside of his right arm, a small round burn on his chest, another on his right thigh, and another on the outside of his right knee where it had been touching Karen's knee. The rescue team pronounced Tim dead at the scene.

Karen was released from the hospital two days later, wrapped in bandages. The dishes, silverware, pots, and pans that she and Tim had bought for their upcoming wedding were still in her car.

After Tim was cremated, Karen and his family carried half of his ashes down to the Missouri River. "We all put our hands on the box, walked to the edge of an obsolete highway bridge, and let it go," his mother says. The other half of his ashes were buried in a family plot. To mark his birthday and other special days, Karen and his family set lighted candles inside waxed bags and float them down the river.

"It was pure destiny," Tim's mother says of his death. "If I hadn't sent him away he wouldn't have lost his fear of lightning. I still have this guilt."

There's a country church along the road near where Tim's parents live. A sign outside proclaims: *"God Is Good All the Time."*

*P*erhaps of all the survivors' stories, Bob Davidson's tale, which has been featured on such TV programs as *Sightings* and *Unsolved Mysteries*, is the most curious. We meet at the survivors' conference in Pigeon Forge and are talking on the patio near the swimming pool of the MainStay Suites. Davidson is sixty-five years old. His sunglasses rest on the top of his head, and he is wearing khaki shorts and white socks. He frowns when he tries to remember something. He often pauses, not from uncertainty but from a manner of speaking that is definite and doesn't lend itself to further questions.

Davidson was born in Findlay, Ohio, about three weeks before the Japanese attack on Pearl Harbor. His father owned a grocery store and, Davidson says, "I was just a typical kid of the time." He quit high school in the eleventh grade and joined the navy, where he served as a machinist's mate on the *Franklin D. Roosevelt* aircraft carrier. He spent four years in the North Atlantic and particularly remembers speeding to Cuba during the missile crisis in 1962. After he left the navy, Davidson took odd jobs, making tires, working in a machine shop, and finally joining General Motors, where he worked for sixteen years before he was struck by lightning.

One early afternoon in June 1980, Davidson and his wife were on his motorcycle heading to Indianapolis on Interstate 74 to visit her sister, who was having a baby. It had started to rain and Davidson pulled off the road. His wife had gotten off the bike and Davidson was dismounting, his left foot on the ground and right leg still on the seat, when a thunderbolt hit him on the shoulder, sending him flying and knocking him out. A physician driving by at the time immediately called a rescue squad and tried to give Davidson CPR, but he couldn't restart his heart.

"I was there, but I was dead," says Davidson laconically. "The rescuers told me all this. The doctor worked on me, and then the rescuers, but they couldn't get a heartbeat." Kimberly Cobb, one of the paramedics interviewed on *Unsolved Mysteries*, insisted, "He had no pulse... he should have been dead."

But then a strange thing happened. A woman appeared,

wearing a long black woolen dress and carrying a Bible. The paramedics found her old-fashioned dress odd, especially on such a hot day. As she moved through the crowd that had gathered around Davidson, she yelled out, "Let me touch him." Two police officers tried to restrain her, but she pushed her way through anyway. Still unable to get a pulse, the paramedics decided they had nothing to lose. "She knelt down beside me," Davidson continues, "and placed the Bible and one hand on my chest and the other hand on the ground. She proceeded to lift the Bible to the sky and muttered something that no one could understand." Then she stood up, smiled at the rescuers, and vanished into the crowd.

Within seconds, Davidson had a pulse. Some of the witnesses, including two of the paramedics, say they never saw the mysterious woman in black, but several paramedics, including Marylou Shafer, were certain. "There is no doubt in my mind," Shafer said. "She was there."

A rescue helicopter flew Davidson to the hospital— "more dead than alive"—where he remained in a coma for seven weeks. When he regained consciousness, he had no idea why he was there. Davidson spent a few more weeks in the hospital, learning to walk and trying to talk, but his vocal cords were injured and he couldn't speak again for the next two years.

Scarred by his injuries, Davidson didn't return to GM for nine years, and his wife divorced him. "I thought my life was over," he says. "I thought of suicide." But he remarried and thinks today that he has become a stronger person. "Usually, when I set out to do something, I don't quit until I get it done."

"And what about the woman in black?" I ask him now.

"Well, where this happened, on the hundred-mile marker on Interstate 74, you go one mile across the field and there is an old museum, sort of an old church, and there's a mannequin in that museum. The rescuers said that the mannequin had on a black dress just like that lady was wearing. They never said it was that dress, but it was like that dress. She was there, then she was gone. I think it was an angel. I don't know if it's our loved ones, part of our families, ancestors, who are always looking over people that are left behind. Whether it's an old aunt, relative, grandmother, I have no idea, and everybody I've talked to—religious, clergy— they keep saying, 'I don't know, it ain't possible.' So I guess I'll have to wait 'til I'm gone to find out."

He believes in God and believes in the hereafter, but "as far as going to church and telling me how you interpret the Bible—I ain't really interested."

Davidson looks out across the lawn toward where his wife is walking. "I don't know if God was sending me a message. I hate to say yes and I hate to say no. He left me here and allowed me to come back to life. Maybe it's to help other people. I'm sure if I am supposed to do something He will tell me. I just hope at the time I'll listen."

*S*teve Marshburn, Sr., is a balding man with glasses, who speaks in a soft Southern drawl, weighing his words carefully. Seriously injured himself by lightning, he has probably spent more time with more survivors than anyone else on the planet, and he has thought deeply about the mysteries and human consequences of a lightning strike for its victims.

His lightning story began in the weeks prior to

Thanksgiving 1969. In those days before direct deposits and electronic withdrawals, workers from the fabric mills, and civilian employees, marines, and navy personnel from the military bases along the North Carolina coast brought their Friday paychecks to the First Citizens Bank and Trust Company on Main Street in Swansboro, built on the former site of an Algonquin Indian village along the White Oak River. The customers waited patiently in long lines that spilled out the front door and down the sidewalk, their progress measured in inches as the tellers inquired about the families of everyone they served.

Marshburn had started working for the bank two years earlier. That afternoon, as the lines grew longer, he knew he should open another teller window. All of the windows were already open except "the one," the station that tellers disliked the most, which was stuck in a corner near the drive-up window with barely enough room to sit on the metal stool. Marshburn signed off for the funds to open the window and slid into the tiny space. His first customer was a friend, and as he unzipped the bag with the large cash deposit the man had handed him, the two started talking amiably. In his right hand, Marshburn was holding a teller's metal date stamp.

Then, from the blue sky above, a lightning bolt from a storm centered over the town of Bogue, ten miles away, struck the drive-up window with a deafening sound. The teller at the window had just stepped away to get a check approved, and, as was customary, she had turned the speaker away from the window. It was pointing straight toward Marshburn's spine. When the bolt struck the drive-up window and traveled through the speaker and into Marshburn's back, "It felt as if someone had hit me with a baseball bat," he recalls. The pain was excruciating, and the left side of his head felt as though it had been split wide open.

Marshburn knew something was terribly wrong. He was semiconscious, but he regained his composure after about thirty minutes. He hadn't even been knocked off the stool.

The next day, Marshburn hurt all over and had an unbearable headache. With his wife's help, he made it up the steps and into a doctor's office. When the doctor checked Marshburn's reflexes and his throat, as if he were looking for the flu, he said he really didn't know what to do but referred him to a friend who was a neurologist in Wilmington, North Carolina, about fifty-five miles away. "I didn't know at the time," Marshburn says in his soft, precise drawl, "that the one split second that lightning impacted my body, entered it, and did its thing, would turn out to be a lifetime of agony."

Despite Marshburn's difficulties, the bank insisted that he return immediately to work. Fearing that somehow business would suffer, the bank never mentioned the strike. People in the town didn't talk about it either. "It was a tiny fishing village, a clannish place back then," says Marshburn. "They were afraid of their reputation." It was as if the incident had never happened.

He returned to his job a few days later, still suffering from his injuries. At the time he was struck, all Steve Marshburn knew about lightning was that one of his cousins had been killed a few years earlier while clamming at Sneads Ferry, North Carolina. Prior to the strike, Marshburn was a laid-back, easygoing person, but during the months and years that followed, he suffered a number of health problems, including chest pains, grand mal seizures, bladder and bowel difficulties, joint stiffness, hearing deficits, internal burns, body cramps, a weakened immune system, muscle spasms, and brain damage. He gained fifty pounds, became moody, depressed, sometimes suicidal, quick-tempered, and phobic in crowds. He went from doctor to doctor—no one knew how to treat him.

Fifteen years passed, and as his health problems grew worse and more complicated, Marshburn came to understand that there was "no doctor on earth who knew what to do for me." Then he was referred to a rheumatologist in Pollocksville, about twenty-five miles away from his hometown of Jacksonville, North Carolina. After only a few minutes, the doctor realized that Marshburn was having trouble forming words and began to complete his sentences for him. When he told Marshburn that he was suffering from "lightning syndrome," Marshburn says, "it felt as if God had opened a window to heaven and heard our pleas."

Marshburn had known adversity before. Born in 1944, he grew up in the small towns near Jacksonville and lived without indoor plumbing until he was nineteen. Of Cherokee and Irish–French descent, Marshburn's father was a Pentecostal minister. "We were always raised in a religious home," he says. "We were taught to respect the Lord and read our Bibles and know that what my dad said was for our own good."

Marshburn had wanted to be a college English professor and had been accepted at East Carolina University, but a temporary job at Sears stretched into three months and then three years. By the time he accepted a permanent position at the First Citizens Bank and Trust Company, he and his young wife, Joyce, had had a baby who was afflicted with cystic fibrosis. "I was only twenty-five years old, with a new wife and baby. I was beginning a new career. It seemed like the powers that be were working on us, testing our strength."

The calamities continued. When he and his wife had another child, they discovered that both children had glomerulonephritis, a renal disease characterized by inflammation of the small blood vessels in the kidneys, and had to be hospitalized. Marshburn's wife had breast surgery for what the surgeon thought was cancer. Their son-in-law, Kevin Thompson, died at age thirty-four from astrocytoma, a type of brain cancer, in 2004. Marshburn's brother, sister-in-law, and niece were all killed in a plane crash. Marshburn was in two automobile accidents. A cousin of his wife's, to whom he was close, was killed when lightning made contact with the springs of an old bed he was sitting on while talking on a land phone, during a thunderstorm.

During his trials, Marshburn stayed at First Citizens Bank and Trust for nineteen years, becoming a mortgage specialist, until a doctor told him one day that he should retire. Marshburn wasn't surprised. "My memory was getting worse. A customer would come in to finalize a loan and I would forget that I had drawn the papers up one day before." Bank employees mocked him for being stooped over while racked with pain and wearing a back brace. He overheard conversations accusing him of faking his injuries.

Since the lightning strike, Marshburn has undergone thirty-seven operations, and every day he is in pain. How much suffering can one human take?

Two decades after he was struck, unable to work at the bank any longer, Steve Marshburn didn't know what to do with the rest of his life. When his wife, Joyce, always at his side, happened to mention to her doctor that her husband was hanging around the house, becoming depressed and obsessing about his injuries, the doctor suggested he find a meaningful project.

On February 22, 1989, Lightning Strike and Electric Shock Victims was formed. ("Victims" has since been dropped, in favor of "Survivors.") The mission of the group is to act as a resource for information; to encourage family members to better understand the symptoms associated with lightning and electrical shocks; to sustain a network of support; and to provide an avenue for members to express their feelings, fears, and emotions.

"It took off," Marshburn says proudly. "It just took off." At the time there was simply no printed information available for survivors. "There was nothing, zero. So the organization began making a difference all of a sudden." (Today it has about 1,500 members.) Marshburn improved mentally and physically. Life became more fulfilling than he could have ever imagined. "I am convinced there was a reason I was struck by lightning—to help others.

"I'm sixty-three now and it's been a way of life," he says. He is thankful that he is not dependent on someone else to take care of him. "I can do most things. I can even drive again." His children are healthy. His son works as an electrician in Oklahoma—"An electrician, mind you." His daughter works as an administrator for a county school system.

After the lightning strike, his faith became stronger. "I knew I had to put my trust in the Lord. I don't see what has happened as punishment. All of us have erred and sinned.

But I don't know of anything I did so drastic that I would be punished so harshly."

Over the years, Marshburn and the organization have aided hundreds of survivors and their families. "You'll see two, three, or four of these people who come to the lightning strike survivors' annual meeting who are near suicidal," observes Dr. Cooper, "who are so depressed they don't know where to go, who may be self-medicating with alcohol. They are not medically sophisticated; they may not have any insurance. They start seeing they're not crazy; a lot of them feel so isolated they feel they are crazy. And they feel it's all their fault the way they are. They feel guilty, depressed, angry, frustrated, isolated. One of the main things they get in the group is support from other people who are there with similar stories. They start seeing a way out of this. They start seeing hope."

Though Marshburn has probably prevented twenty or more suicides, "I don't want any pats on the shoulder," he says modestly. "The pats on the shoulders should go to the people who have healed themselves and helped others."

He now spends most of his time answering questions from survivors and their families. His organization has published the three volumes of *Life After Shock*, enabling survivors to tell their stories; ten self-help books; and several brochures. At the meeting in Pigeon Forge, one woman from Texas tells the group that her husband didn't reveal for fifteen years that he had been struck and only admitted the truth after she threatened to leave him. A single mother from Ohio tells of losing her job after being hit by lightning and of worrying about having enough money to feed her two children. Participants at the conference also describe the bureaucratic obstacles they face, such as applying for Social Security disability payments and health-insurance reimbursements.

"Had I been the potter and He the clay, I could not have set the stage for my life any better," Marshburn says of his work with lightning and electric shock survivors. "When my mind is on helping other people, I'm not hurting."

*F*or many lightning survivors, the future eventually takes precedence over the past. With much pain and effort, they have worked through their confusion and frustration, their anger and depression. What has happened can be determined more or less. Why it happened is left to conjecture and faith. But as many survivors enter the acceptance stage of their recovery, they often feel a sense of obligation, a responsibility to act. They feel they are expected to do something meaningful, something that justifies their having been spared. They seek to restore control over their lives, often with renewed strength and courage.

People who have worked closely with lightning survivors believe that being struck is a way for survivors to examine their lives and become better human beings. When asked how they've changed, almost every survivor says they seek to be more patient, tolerant, and kind, and to place less importance on material things and more to the worlds of faith and the spirit. They almost always believe that the lightning strike was not punishment, whether divine or otherwise, for past deeds. The challenge of the future is what matters.

At the MainStay Suites in Pigeon Forge, Tennessee, the man who has stared down bears has a wistful look in his blue eyes. "I'm a stargazer," Griz says. "I love clear nights when all the stars are out. Some nights I'll just sit there and think—why did You let this happen?

"I still feel there's something that I'm supposed to be doing that I haven't found yet. I feel that I'm supposed to be doing something else but I don't know what it is."

"It's not a big mission survivors have," says Dr. Cooper, "to go out and get a Nobel Prize or something. It's a mission of just leading a good life. And you never know how people to whom you said just the right thing at the right time—how it changed their life."

I recall the lucky scout Mark Evans, who served as a church missionary in Peru after his injury and believes that "God stepped in after the strike and allowed me to live so that I could grow and bless the lives of others." And Gwynn Blair, a fifty-seven-year-old former Florida state trooper and self-described "simple-minded ol' farm boy," who told me, "God works through so many different ways. It took being hit by lightning to get me where I need to be. Bad things happen to the just as well as the unjust. And I think God lets it happen to see how we will handle it. Character shows."

Ed Ellickson is also philosophical. "This whole chain of events has a purpose, even though I don't necessarily understand it. We have faith that this will turn out okay given enough time. Now, maybe that's a blind faith, maybe that's a foolish faith, but to me there's a purpose to this. I couldn't tell you what it's for. The chapter hasn't been written."

*A*s a thunderstorm threatens outside the hotel lobby where we are talking, Marshburn is getting restless, worrying how other survivors will react to the storm. It seems as good a time as any. "What are your thoughts about fate?" I ask.

He looks at me with an intensity in his eyes. "Sometimes it's being in the wrong place at the wrong time. Twenty-one people are leaning on a chain-link fence in a storm at a softball game in Fayetteville, North Carolina, and they all get mowed down by lightning. Now, that was not fate. To me that was not using common sense."

Then he leans forward as if to reveal a secret. "I'm going to tell you exactly how I feel," he says in almost a whisper. "This is my own gut feeling and it is not the official position of the organization. I firmly believe that persons who have been injured—I'm not saying it's by God, I'm not saying it's by fate, I'm not saying it was a freak thing—are handpicked. They are the cream of the crop. They get better and teach other people who have been injured. They have a purpose. I just think that maybe God knows who will answer the call to help other people. I can say that people who have been injured are people who help other people.

"You know, life is not a promise to be happy," says Marshburn. "I look back at the apostle Paul and Martin Luther. Did they help people after they were struck? Sure they did. I feel that maybe I and the others who were injured are part of that group of helpers."

An injured climber dangles below the helicopter as pilot
Laurence Perry rescues him off the Grand Teton a few hours
after the lightning strike on July 26, 2003.
(Photo by Leo Larson, National Park Service)

THE MOUNTAIN CLIMBER:

THE DESCENT

I have set before you life and death, blessing and cursing: therefore choose life, that both thou and thy seed may live.
—DEUTERONOMY 30:19

The rangers have worked for hours to transfer Rod up the mountain and onto a litter. Now, with Rod barely responsive, daylight is fading and with it the hope of using the short haul to take him to safety.

"Clipped and ready," Holm radios to Perry at 8:57. "You ready to raise?"

"Lifting," Perry replies. The litter rises, silhouetted against the darkening sky. Holm moves closer to his patient to monitor his airway in case he vomits, which could cause him to suffocate. He is almost sitting underneath the injured man on his harness, his rope attached to the helicopter. "We're flying," he shouts in the wind to Rod, his face six inches away. "We're going to set you down in the meadow. It's almost over."

At 9:05, Perry brings Rod's litter down gently in Lupine Meadows. A flight nurse, a paramedic, an emergency-room doctor, and other rangers are waiting to transfer Rod to the hospital. He is the last survivor off the mountain.

It's almost pumpkin time, when the rescue helicopters will be grounded for the night. Rod and his close friend Jake Bancroft are placed aboard an Air Idaho helicopter, permitted to fly at night, that takes them to the Eastern Idaho Regional Medical Center in Idaho Falls.

Minutes before, Perry's helicopter has retrieved Erica Summers's body from the upper rescue scene and returned with her to Lupine Meadows.

The uninjured climbers have made their way down Garnet Canyon. The rangers break down their anchor systems and descend by headlamp along the Exum Ridge and then down to the Lower Saddle, leaving their gear on the ledges to take out the next day. As they rappel through the darkness, past familiar landmarks, they alternate between silence and occasional conversation.

Some think to themselves that though there have been rescues in more difficult terrain and weather, what set this mission apart is the number of critically injured people involved, in several locations, high on the mountain, in unsettled weather, only five hours before nightfall. Six survivors have been flown off the upper reaches of the Grand in the span of three hours. Tom Kimbrough, a thirty-year veteran ranger and the most senior Jenny Lake ranger who has helped with the operation, remarks, "This might be the most spectacular rescue in the history of American mountaineering."

The rangers know that their success is due partly to luck. Things would have been different if the clouds had thickened or a storm from the west had moved in. Instead of being airlifted, Rod Liberal would have spent the night out-doors—and likely died. The rest of the injured would have remained in below-freezing temperatures, and it's not clear if they would have survived.

At about 9:23, pumpkin hour, when all contract helicop-ters have to conclude operations, Perry finally stops flying. He has flown ten round-trips, rescuing the climbers and in-serting the rangers and their supplies. It's the highest eleva-tion rescue he has ever flown.

By midnight, all of the rangers and climbers are off the Exum Ridge route and at the Lower Saddle. For the other survivors, one of the worst days of their lives is behind them. But for Rod Liberal, the horror is just beginning.

———

*A*t the hospital in Idaho Falls, Rod was misdiagnosed as having burns on fifty-two percent of his body, but when he was immediately flown to the Intermountain Burn Center at the University of Utah in Salt Lake City, it was correctly de-termined that he had burns on only thirteen percent of his body. In addition, he had kidney failure and needed to re-ceive constant dialysis. One of his lungs was filled with fluid. His right leg was paralyzed. He had pancreatitis.

Doctors put him into a medically induced coma. "It was nightmarish and ghoulish," he recalls. "I had hallucina-tions—hospital rooms that were infested with monsters and horrible creatures. I had dreams of being in a hospital that had a conveyor system with bodies of mutilated people. They were floating around all over the place. In another dream Jody was pushing me in a wheelchair through a mall. I was really old and she was telling me, 'This is your last day on earth. What do you want to do?' "

Before he was put into the coma, Rod first learned that Erica Summers had died. "I knew that somebody had died on the mountain. When I heard it on the radio on the mountain, I don't remember worrying who it was. I learned it sometime after I had arrived at the burn center. The first day in the hospital, I told Jody that Erica had died, and I cried uncontrollably."

After three weeks in the coma, Rod woke up and spent forty-three days in the ICU before being transferred to an intensive-rehabilitation facility in Salt Lake City. He had to learn to sit up, to walk again. "I had spent six weeks in the hospital with no window. I finally get to the rehab facility and I've got this big window right next to my bed. Well, that night there was a nasty storm, and they had to move the bed all the way over to the other side, close to the wall. I don't think I slept at all. That was the extreme of my fear of lightning. Today it's not so bad."

On September 26, exactly two months after the lightning strike and two days before his twenty-eighth birthday, Rod Liberal went home. "I was extremely depressed. I was taking drugs, antidepressants. I went to work not long after I got out of rehab, but I was a mess. I couldn't focus on anything. I couldn't remember anything. I was completely unproductive. The company kept my paychecks coming, but I was not helping the company. Jody and I finally decided to move back to Salt Lake City."

We are sitting in his kitchen in Layton, a Salt Lake City

suburb. It is a sunny day and the blinds are partially closed. The low green hills of the Wasatch Mountains are only a few miles away. I'm struck by the contrast between the unspoiled Tetons and the Kennecott Copper Mine, the largest mine on earth, which has despoiled the surrounding mountains near Salt Lake City, and by the contrast between the Tetons' fresh mountain air and the sulfurous odor of the Great Salt Lake, America's Dead Sea, which often wafts through Salt Lake City.

Seed pods from cottonwood trees fall lazily like tiny parachutes outside Rod's modest split-level house. It is sparsely furnished. His wife, Jody, and son Kai, who is now four years old, walk in and out of the kitchen. Eight-month-old Brennan starts to cry in the living room, then stops.

As Rod relates the story of his ordeal, Jody interjects when Rod doesn't remember. "I've always had an awful memory," he says. "It's gotten back to where it was before the accident—bad."

Rod still has numbness in his right foot. He's suffered muscle and nerve damage in his right leg. He has hypersensitivity in his left leg. Some scars. He can still play hockey and skate, but his legs buckle every once in a while. "Overall I'm okay," he says.

Erica Summers was buried on July 31, 2003—the day before her fifth wedding anniversary. Clinton told their children that their mother was in heaven. "These things happen for a reason," he says.

Of the other injured climbers, Jake Bancroft suffered a severe concussion, and two other climbers suffered a number of serious injuries, including broken bones, ripped tendons, and severe burns. Clinton Summers suffered painful burns and underwent treatment to remove damaged skin. The climbers also had to endure criticism from the climbing community, especially for starting their ascent so late in the day and for not paying attention to warnings about afternoon thunderstorms.

Six months after the accident, Rod took his family to see a film that had been made about the rangers before the accident. "Many of the rangers were there, including Craig Holm," he remembers. "He's my hero."

Holm has since "quit the best job in the world" as a ranger to accept a more financially secure position with the Rocky Mountain Fire Department in Boulder, Colorado, teaching rescue skills. But every year he goes back to the Tetons—"my favorite place ever."

While watching the movie, Rod thought he recognized one of the pilots. "I realized it was Laurence Perry and he was sitting in front of me. After the movie I tapped him on the shoulder and I said, 'Perry,' and I'm about to introduce myself when his eyes just light up. He has a real cool British accent. He got really emotional. 'Look, honey,' he said to his wife. 'It's Rod Liberal.'

"The rangers saved my life and I owe them something," Rod says. "But they're so humble. They just say it's their job."

Blond-haired Kai crawls up on his father's lap. "I'm definitely more aware of how fragile we are. I did extreme sports all my life. No real thought of the consequences. Then I have a new kid and I'm up in the Tetons?"

Two months after the accident, Robert Thomas scaled

the Grand and built a cairn where Erica and Clinton had been sitting. He removed a five-pound piece of granite, which he had engraved in memory of Erica and the climbers who survived that day on Friction Pitch.

The whole group, except for Rod, returned to the site on the first anniversary. On the second anniversary, Rod made the climb with Jake and two others who had not been on the first climb. "I suppose I needed a sense of accomplishment and closure. When I went back we did things differently. We left at two in the morning."

When Rod reached Friction Pitch, he recalls, he "got pretty emotional. I saw where Erica died. It was pretty intense being up there. But reaching the summit was the best feeling in the world—touching the marker. I think it was the same feeling as if I had made it the first time. I have climbed the Grand Teton. For a climber, that's a pretty neat thing to do. We didn't stay on the summit too long, maybe twenty or thirty minutes. Then we headed back. We were in the parking lot by ten at night."

On the walls of Rod's house hang several wooden boards with inscriptions painted by Jody. One board reads: "Life is not measured by the number of breaths we take, but by the moments that take our breath away."

"After the accident I couldn't stop thinking about death," Rod says. "I could go anytime, whether driving down the road, sitting on a porch, just doing anything. I think that's how it affected me. I don't worry too much about little problems like bills not getting paid. When things like that come up, it's no big deal.

"After we moved in to this house, I became very active in

the Church of Latter-day Saints. I started taking Kai with me, and I loved it. We got to know the Mormon missionaries, the guys with the suits, walking around with the name tags.

"Jody had discussions with the missionaries, and she converted and was baptized. Then we went to church as a family. It was great. After a certain period of time we were able to attend sacred ceremonies in the temple, but not long after, I started having questions about church history and certain doctrines, and it didn't take too long for me to start discounting claims. To make a long story short, I changed, and I guess I consider myself an atheist today. But at the time that I was injured on the mountain, I honestly think I had some communication with the High Being. I mean, it was a real spiritual experience for me. I am begging somebody to let me live, and it was a conscious decision to do that, but I think now it was based on what I believed then.

"I think what brought me back to the Church was the belief that I was given a second chance. I really believed that there was a purpose for me being here. And I heard that over and over again. People I talked to said you definitely have work to do here. God wants you here for a reason. You wouldn't think that way if you didn't believe in anything. You wouldn't think it happened just because it happened— that there's no fate involved, no higher purpose or anything like that. But I don't believe that now.

"Sometimes I ask, why didn't Erica have the same chance I did? I mean, she had two kids at the time. You can argue it a hundred ways. Maybe it was her time. Maybe she has a job to do somewhere else."

In fact, it turns out that lightning struck twice that July day on the Grand. One bolt hit Rod's group. An earlier bolt struck harmlessly farther away. What if their order had been reversed and the climbers had been kept away from danger?

Was it fated that Erica Summers would die and Rod Liberal live? Or was it simply random chance, like the spinning ball on the roulette wheel?

EPILOGUE

*O*n a dry mesa of scrub pines high in
the mountains of southern New Mexico, beneath an endless
sky, where unseen coyotes howl and Apaches once rode, a
perplexing and mysterious installation protrudes from an
open field. Considered one of the twentieth century's most
significant works of land art, Walter De Maria's *Lightning
Field*, which opened in 1977, was built over a period of ten
years in secrecy at a cost of half a million dollars. The field's
creator is a "conceptual" artist whose other notable works
include a metal shaft sunk one kilometer into the earth in
Kassel, Germany, and a four-mile-long, six-foot-wide walk-
way cutting across an isolated stretch of the Nevada desert.

The *Lightning Field* is composed of four hundred pol-
ished stainless-steel poles with pointed tips, each two inches
in diameter and approximately twenty feet high. The poles
are embedded permanently in the earth, at intervals of 220
feet, in a grid covering a rectangular area that measures one

mile on one side and one kilometer on the other. When lightning strikes a pole, it has been known to jump erratically to the other poles, like the notes in a symphony bouncing from one instrument to another. Despite the work's provocative name, the effect of lightning storms on the field is only an incidental consideration. The site's official guide states: "It is important to note that as a work of art, the *Lightning Field* does not depend upon the occurrence of lightning but responds to many more subtle conditions of its environment."

When I first visited the field, about twenty-five years ago, I was struck by its remote desolation and wondered why anyone would construct something so inaccessible. Visitors must drive for forty-five minutes over a dirt road, often clogged with mud, from the tiny town of Quemado, New Mexico, which itself is almost three hours from Albuquerque. Nonetheless, the artwork engenders a strong emotional response. Richard McCord, a former editor of the *Santa Fe Reporter*, captures the dramatic effect of the place during a lightning strike: "Braced and in place, I watched the yellow light of day lodge first in the western peaks, then sweep down toward where I waited. The fire ignited spontaneously at the pointed tips of the poles. The strains of dervish music were almost audible in the air, and surrounded by those pulsing poles of light, the unleashed imagination could conjure up a sense of being 'present at the creation'—as earlier awestruck observers in southern New Mexico had felt upon a much grimmer occasion: the detonation of the first atomic bomb."

*I*t's a long distance in many ways from a barren mesa in New Mexico to a movie set in Rome, and yet the environment of the *Lightning Field* allows one to imagine how the ancients must have viewed the awesome thunderbolt and brings me to a final survivor's story.

James Caviezel, the actor who portrayed Jesus in Mel Gibson's film *The Passion of the Christ*, was struck by lightning in 2003 while filming the crucifixion scene for the movie in Rome. "When I was hit by lightning," he recalled, "it was the one day I didn't have communion. We always had Mass and I always received communion, but on that one day the priest ran out of hosts. I was up there on the cross and I was hit."

What can we take away from Caviezel's extraordinary experience? A religious person might see it as a sign. But a sign of what? Not to portray Jesus in a certain way? Not to dare to portray Jesus at all? Was it a warning that some dire event was going to happen or that the actor should change the direction of his own life? Since Caviezel wasn't hurt, was it some form of divine favor?

Roy Sullivan, the ranger in the *Guinness World Records* for being struck seven times, believed, "Some people are allergic to flowers, but I'm allergic to lightning. I think I must have some chemical or mineral in my body that attracts lightning."

Carl Mize, the ex–rodeo rider who was hit six times, says, "My boss who runs the power plant at Oklahoma University believes there's something in my body that attracts lightning."

Linda Cooper, no relation to Dr. Cooper, who was struck four times, claims, "The magnetic force of my body pulls lightning toward me. I may be totally wrong, but you know when you magnetize something it remains magnetized until you demagnetize it. Well, guess what? I'm not demagnetized. Every time I go through an airport security device, I set it off."

What is known is that, for whatever reason, a few people like Sullivan, Mize, and Cooper have repeatedly been victims of lightning strikes and yet, ironically, it has never caused them fatal injury. As far as can be determined, no one who was struck once and hit a second or third time has ever died from the subsequent strikes.

Doctors and lightning experts, including Dr. Mary Ann Cooper, are highly skeptical that an individual's physical characteristics can attract or repel lightning or that people who have been struck can alter electronic devices. Furthermore, the physicians emphasize that survivors often embellish their experiences, don't really remember what happened, make up stories, or rely on the accounts of others.

I've let the survivors speak for themselves. Their experiences, symptoms, and questions cannot be ignored. No long-term studies have been conducted on lightning survivors, and modern medicine still knows very little about what happens after a person is hit. "We can't even define what has happened. We can only describe it," Dr. Cooper observes.

And yet progress has been made. "Some of the research we've done has had an impact," she admits. "We have helped people. We have prevented injury. We helped to decrease the mortality rate by forty percent in the 1990s, and that means that two hundred or two hundred fifty people who would have been killed in the U.S. are alive."

One of the most promising areas of research with lightning victims is the use of functional MRI, which creates a map of brain activity "in real time" and highlights areas involved in language, motor control, and sensory activity. Tests using this MRI have revealed that lightning survivors have shown either more or less activity in parts of their brain than is normal.

While the medical results of lightning strikes remain difficult to describe and define, so, too, are the spiritual and psychological effects. "Perhaps faith and spirituality arise where the limitations of medicine leave off," says Dr. Cooper.

The survivors I interviewed represent a broad cross section of individuals: men, women, black, white, young, old, rich, poor, Catholic, Jewish, Mormon, Protestant, and non-believers. They were an average group of people, surely neither morally superior nor inferior to their neighbors before or after they were struck. But clearly they are unique. They were singularly affected, without explanation or cause, by a life-threatening force of nature, and then they were spared.

As to the "thousands of little facts and anecdotes and testimonies" that Brother Juniper gathers in *The Bridge of San Luis Rey* to try to learn "why God had settled upon that person and upon that day for His demonstration of wisdom," such an investigation might yield more clues about human existence if it examined how the lives of lightning victims changed after their accidents rather than what paths led to their tragedy. Being struck by lightning is not about what survivors did in the past but about what they will do in the future.

The majority of survivors believe that the thunder-bolt that felled them was not punishment for their past deeds but rather a life-changing moment that obliges them to reevaluate their lives. Such conversions may be met with skepticism but deserve to be credited if they are subsequently followed by action.

In the end, my exploration of the mysteries of lightning led to an increased fear and respect for its random force and awe of its power to transform those it touches. The resilience of the human spirit should not be underestimated.

"People who have lightning injuries have taught me so much," says Dr. Cooper. "They've taught me so much about belief and about faith and about triumphing over the problems that have been set in their way.

"You'll find that most of us in lightning work are fairly religious," she tells me. "I think that we've been put on this earth to help others. Whether that means teaching somebody in the emergency department to take care of stress or whether it's my lightning work or whether it's teaching third- and fourth-graders in Sunday school.

"People who were struck say, 'Hey, I should have died. I didn't die. There must be a reason for me to be here.' I think more of us need to think about such things. We might have a much better world if people thought about why they are here on earth and what God might have in mind for them instead of what they have in mind for themselves and what they're entitled to.

"What guided me into doing the things that I'm doing? Where has God directed me? I realize that I didn't put this all together on my own. So I give Him all the credit.

"Sometimes, when there is a big storm, I sit with my children on our front porch, and we shell peanuts and watch the lightning. I think it's one of the most beautiful things God has given us. It's just spectacular."

BIBLIOGRAPHY

BOOKS AND OTHER SOURCES

Andrews, Christopher J., and Mary Ann Cooper, Mat Darveniza, David Mackerras. *Lightning Injuries: Electrical, Medical, and Legal Aspects.* Boca Raton FL, Ann Arbor MI, London, and Tokyo: CRC Press, 1991.

Aristophanes. *The Clouds.* The Internet Classics Archive. classics.mit.edu/Aristophanes/clouds.html.

Austin, Brian. *Schonland: Scientist and Soldier.* Bristol, England, and Philadelphia PA: Institute of Physics Publishing, 2001.

Blackmore, Susan. *Dying to Live: Near-Death Experiences.* Buffalo NY: Prometheus Books, 1993.

Brecht, Martin. *Martin Luther: His Road to Reformation, 1483–1521.* Tr. by James L. Schaaf. Philadelphia PA: Fortress Press, 1985.

Bruchac, Joseph. *Flying with the Eagle, Racing the Great Bear: Stories from Native North America.* n.c.: Troll Medallion, 1993.

Cerveny, Randy. *Freaks of the Storm: The World's Strangest True Weather Stories.* New York: Thunder's Mouth Press, 2005.

Cheney, Margaret. *Tesla: Man Out of Time*. New York: Dell Publishing, 1981.

Cohen, I. Bernard. *Benjamin Franklin's Science*. Cambridge MA and London: Harvard University Press, 1990.

Delbourgo, James. *A Most Amazing Scene of Wonders: Electricity and Enlightenment in Early America*. Cambridge MA and London: Harvard University Press, 2006.

De Villiers, Marq. *Windswept: The Story of Wind and Weather*. New York: Walker and Company, 2006.

Dray, Philip. *Stealing God's Thunder: Benjamin Franklin's Lightning Rod and the Invention of America*. New York: Random House, 2005.

Eagleman, Joe R. *Severe and Unusual Weather*. Lenexa KS: Trimedia Publishing Company, 1990.

Edwards, Jonathan. *The Works of President Edwards*, Vol. 1. New York: S. Converse, 1829.

Ehrlich, Gretel. *A Match to the Heart*. New York and London: Penguin, 1995.

Franklin, Benjamin. *The Autobiography of Benjamin Franklin*. New Haven CT: Yale University Press, 1964.

Frazer, James G. *The Golden Bough*. New York: The Macmillan Company, 1967.

Freier, George D. *The Wonder of Weather*. New York: Grammercy Books, 1999.

Hamilton, Edith. *Mythology*. Boston: Little, Brown and Company, 1942.

Hammond, John Winthrop. *Charles Proteus Steinmetz: A Biography*. New York and London: The Century Co., 1924.

Herodotus. *Herodotus*. Tr. by Reverend William Beloe. Philadelphia: Thomas Wardle, 1840.

Hesiod. *The Homeric Hymns and Homerica*. Tr. by Hugh G. Evelyn-White. Cambridge MA and London: Harvard University Press, 1977.

Hindle, Brooke. *The Pursuit of Science in Revolutionary America, 1735–1789*. Chapel Hill NC: University of North Carolina Press, 1956.

Isaacson, Walter. *Benjamin Franklin: An American Life*. New York, London, Toronto, and Sydney: Simon and Schuster, 2003.

Krakauer, Jon. *Into Thin Air*. New York: Anchor Books, 1999.

Kübler-Ross, Elisabeth. *On Death and Dying*. New York: Scribner, 1997.

Lane, Frank W. *The Violent Earth*. Topsfield MA: Salem House, 1986.

Lavine, Sigmund A. *Steinmetz: Maker of Lightning*. New York: Dodd, Mead and Company, 1955.

Life After Shock: Members Tell Their Stories. Jacksonville NC: Lightning Strike & Electric Shock Survivors International, Inc., Vol. I, 1996; Vol. II, 2000; Vol. III, 2007.

Marsden, George. *Jonathan Edwards: A Life*. New Haven CT and London: Yale University Press, 2003.

Melville, Herman. *Moby Dick*. New York: The Modern Library, 1950.

Miller, Floyd. *The Man Who Tamed Lightning: Charles Proteus Steinmetz*. New York: Scholastic Book Services, 1965.

Moody, Raymond A., Jr. *Life After Life*. Harrisburg PA: Stackpole Books, 1976.

Ortenburger, Leigh N., and Reynold G. Jackson. *A Climber's Guide to the Teton Range*. Seattle WA: Mountaineers Books, 1996.

Plato. *Phaedrus*. Tr. by Benjamin Jowett. http://ccat.sas.upenn.edu/jod/texts/phaedrus.html.

Pretor-Pinney, Gavin. *The Cloudspotter's Guide: The Science, History, and Culture of Clouds*. New York: A Perigee Book, 2006.

Priestley, Joseph, *The History and Present State of Electricity with Original Experiments*, Vol. 1, 3rd Edition. London: C. Bathurst and T. Lowndes, 1775.

Rakov, Vladimir A., and Martin A. Uman. *Lightning: Physics and Effects*. Cambridge, New York, Melbourne, and Cape Town: Cambridge University Press, 2003.

Renner, Jeff. *Lightning Strikes: Staying Safe Under Stormy Skies*. Seattle WA: Mountaineers Books, 2002.

Reuther, David, and John Thorn, eds. *The Armchair Mountaineer.* Birmingham AL: Menasha Ridge Press, 1998.

Riskin, Jessica. *Science in the Age of Sensibility: The Sentimental Empiricists of the French Enlightenment.* Chicago and London: The University of Chicago Press, 2002.

Rowling, J. K. *Harry Potter and the Sorcerer's Stone.* New York: Arthur A. Levine Books, 1998.

Schonland, B. F. J. *The Flight of the Thunderbolts.* Oxford: The Clarendon Press, 1950.

Shakespeare, William. *King Lear. The Complete Plays and Poems of William Shakespeare*, eds. William Allan Neilson and Charles Jarvis Hill. Boston, New York, and Chicago: Houghton Mifflin, 1942.

Stubbs, Dacre. *Prehistoric Art of Australia.* New York: Charles Scribner's Sons, 1974.

Suetonius. *The Twelve Caesars.* New York: Penguin Classics, 1985.

Tolstoy, Leo. *Anna Karenina.* New York: Modern Library, 2000.

Uman, Martin A. *All About Lightning.* Mineola NY: Dover Publications, 1987.

Van Doren, Carl, ed. *Benjamin Franklin and Jonathan Edwards: Selections from Their Writings.* New York, Chicago, and Boston: Charles Scribner's Sons, 1920.

Viemeister, Peter E. *The Lightning Book: The Nature of Lightning and How to Protect Yourself from It.* Cambridge MA and London: The MIT Press, 1972.

White, Andrew Dickson. *A History of the Warfare of Science with Theology in Christendom.* New York: D. Appleton and Company, 1898.

Wilder, Thornton. *The Bridge of San Luis Rey.* New York: HarperCollins, 2003.

ARTICLES AND OTHER SOURCES

Beilinson, Jerry. "Struck by Lightning on Grand Teton: Behind the Rescue," *National Geographic Adventure*, September 17, 2003, nationalgeographic.com.

Britt, Robert Roy. "Zap! Rockets Trigger Lightning, Scientists Discover X-Rays," http://www.space.com/businesstechnology/technology/rocket_lightning_030130.html.

Bryan, Bobette. "The Lady in Black: A Miracle?" *Underworld Tales Magazine*, 2005, www.underworldtales.com/miracle.

Cherington, Michael. "James Parkinson; Links to Charcot, Lichtenbergh, and Lightning," *Archives of Neurology*, Vol. 61, June 2004.

Cohen, Sharon. "A lightning strike in the Tetons leaves a climber dangling," Associated Press, October 18, 2003.

Cohen, Sharon. " 'Am I going to be OK?' dangling climber asks as rangers finally reach him," Associated Press, October 20, 2003.

Cohen, Sharon. "Helicopter fights winds and rangers scramble," Associated Press, October 18, 2003.

Conley, Brian, and Sina Najafi. "All About Lightning: An Interview with Martin Uman," *Cabinet*, Issue 3, Summer 2001, www.cabinetmagazine.

Cooper, Mary Ann. "Disability, Not Death, Is the Main Problem with Lightning Injury," Abstract from an Address presented at the National Weather Association annual meeting, Oklahoma City, Oklahoma, October 1998, http://www.uic.edu/labs/lightninginjury/Disability.pdf.

Cooper, Mary Ann. "Lightning Injuries," *eMedicine*, emedicine.com/emerg/topic299.htm.

Cooper, Mary Ann. "Lightning Injuries: Prognostic Signs for Death," *Annals of Emergency Medicine*, 9:3, March 1980.

Cooper, Mary Ann. "Lightning Injury Facts: Myths, Miracles, and Mirages," Adapted from *Seminars in Neurology*, Vol. 15, No. 4, December 1995.

Cooper, Mary Ann, and Steve Marshburn. "Lightning Strike and Electric Shock Survivors, International," *NeuroRehabilitation*, 20, 2005.

Crary, David. "Public Safety, Are Scouts Put at Risk," Associated Press, January 15, 2006.

Critchley, Macdonald. "Neurological Effects of Lightning and of Electricity," *The Lancet*, January 13, 1934.

Curran, E. B., R. L. Holle, and R. E. López. "Lightning Casualties and Damages in the United States from 1959 to 1994," *Journal of Climate*, 13, 2000. Data is also available at the National Severe Storms Laboratory Web site: http://www.nssl.noaa.gov/papers/techmemos _NWS-SR-193/techmemo-sr193.html.

Daley, Jason. "Struck," *Outside*, Vol. 30, No. 10, October 2005.

Dollinger, Stephen J. "Lightning-strike disaster among children," *British Journal of Medical Psychology*, 58, 1985.

Dworschak, Manfred. "Lightning Strike Survivors Meet for World Conference," Part 3, Spiegel Online International, www.spiegel.de/international.

Edlich, Richard F., and David B. Drake. "Burns, Lightning Injuries," *eMedicine*, emedicine.com/plastic/topic517.

Foster, J. Todd. "Story of Dooms man who survived 8 lightning strikes might play prime time," *The News Virginian*, January 3, 2003.

French, Thomas, and Sheryl James. "Lightning, Nature's Strike Force," *St. Petersburg Times*, July 23, 1989.

Fuhrmeister, Christian. "The Political Iconography of Lightning." Paper presented at Taming the Electrical Fire: A Conference on the History and Cultural

Meaning of the Lightning Rod. The Bakken Library and Museum, Minneapolis MN, Nov. 4–6, 2002.

Gatewood, Medley O'Keefe, and Richard D. Zane. "Lightning Injuries," *Emergency Medicine Clinics of North America*, 22, 2004.

Haines, Lester. "Canadian iPod user struck by lightning," *The Register*, http://www.theregister.co.uk/2007/07/13 /ipod_lightning_strike.

Haraszti, Zoltan. "Young John Adams on Franklin's Iron Points." *Isis*, Vol. 41, No. 1.

Hetze, Claudia. "Lightning History," www.meteoros.de /light/blitze_he.htm.

Hoadley, David. "Roger Jensen—A Friend," from the WX-Chase List, May 14, 2001, www.onthefront.ws /jensen.htm.

Hoadley, David. "Why Chase Tornadoes?" Stormtrack, www.stormtrack.org/library/chasing/whychas.htm.

Holle, Ronald L. "Activities and Locations of Recreation Deaths and Injuries from Lightning," paper presented at the International Conference on Lightning and Static Electricity, Blackpool, England, 2003.

Holle, Ronald L. "Annual Rates of Lightning Fatalities by Country," paper presented at the International Conference on Lightning and Static Electricity, Paris, France, 2007.

Holle, Ronald L. "Lightning Caused Deaths and Injuries in the Vicinity of Water Bodies and Vehicles," paper presented at the International Conference on Lightning and Static Electricity, Paris, France, 2007.

Holle, Ronald L., and Raúl E. López. "A Comparison of Current Lightning Death Rates in the U.S. with Other Locations and Times," Preprints, International Conference on Lightning and Static Electricity, Blackpool, England, 2003.

Hotz, Lee. "Lightning Strikes Man for the Seventh Time," *The News Virginian*, Summer, 1977.

Johnson, Linda. "Lightning strikes reported by iPod users," Associated Press, July 11, 2007.

Kithill, Richard. "Annual USA Lightning Costs and Losses," National Lightning Safety Institute, http://www.lightningsafety.com/nlsi_lls/nlsi_annual_usa _losses.htm.

Lawlor, Christopher. "Prep player dies after lightning strike," *USA Today*, September 16, 2004.

Loeb, Vernon. "A Killer Bolt Out of the Blue," *Los Angeles Times*, August 16, 2005.

Marshall, Tim. "Roger Jensen: A Storm Chasing Pioneer," www.stormtrack.org/jensen.

Matthews, John. "Fear of Lightning Grounded in Myth," ASU Research, http://researchmag.asu.edu/stories /lightning.html.

McCord, Richard. "First It Reached Out on an Intellectual Plane," *The Santa Fe Reporter*, Vol. 5, No. 38, March 22, 1979.

Melville, Herman. "The Lightning-Rod Man," www.classicshorts.com/stories/tlrm.htm.

Muehlberger, Thomas, Peter M. Vogt, and Andrew M. Munster. "The long-term consequences of lightning injuries," abstract, *Burns*, 27, 2001.

National Park Service, United States Department of the Interior, Case Incident Record, Lightning Strike Fatality/Mountain Rescue and Supplementary Case/Incident Record, July 26, 2003.

National Weather Service. "Updated AMS Recommendations for Lightning Safety—2002," www.lightningsafety.noaa.gov.

Noonan, Peggy. "A Great Moment in the Life of an Artist: Talking with James Caviezel After His Meeting with the Pope," *Opinion Journal*, March 18, 2004.

Parry, Wayne. "Official: Scouts sent campers to tents during storm," phillyBurbs.com.

PBS. "Tesla: Master of Lightning: Life and Legacy, Colorado Springs," http://www.pbs.org/tesla/ll/ll_colspr.html.

Reed, Mary. "Weather Talk," *Weatherwise*, June 1988.

Rosellini, Lynn. " 'We've Been Hit!' " *Reader's Digest*, June 2004.

Sacks, Oliver. "A Bolt from the Blue," *The New Yorker*, July 23, 2007.

Schrope, Mark. "The Bolt Catchers," *Nature*, Vol. 431, September 9, 2004.

Simkins, Charles "Nick." "Can Long-Term Cognitive and Emotional Problems Be Caused by Electric Shock and Lightning Strike Accidents or Is Anything That I Know About Brain Injury Applicable to Electric Shock and Lightning Strike Victims?," paper written for Lightning Strike and Electric Shock Survivors International, n.d., www.lightning-strike.org/DesktopDefault.aspx ?tabid=63.

Thuermer, Angus, Jr., and Rebecca Huntington. "Climbers: Rangers Saved Us," *Jackson Hole News & Guide*, July 30, 2003.

Torres, Brandon. "MCI in the Clouds," *Jems*, Vol. 29, Issue 12, December 2004.

Wachtel, H., Michael Cherington, and Philip Yarnell. "Sub-Visible, Low Thermal, Lightning May Account for Cases of 'Unexplained' Sudden Cardiac Death," presented at the Annual Conference of the Bioelectromagnetics Society, Cancún, Mexico, June 2006.

Weingarten, Paul. "The Tornado Chaser's Vacations Take a Strange Twist," *Chicago Tribune*, June 4, 1980.

Wright, Fred W., Jr. "Florida's Fantastic Fulgurite Find," *Weatherwise*, July/August 1998.

Yarnell, Philip. "Neurorehabilitation of Cerebral Disorders Following Lightning and Electrical Trauma," *NeuroRehabilitation*, Vol. 20, No. 1, 2005.

NOTES

AN AWESOME FLAME
(page number listed)

1 *The stars are the campfires:* Telephone interview with Julie Spotted Eagle Horse, May 29, 2007.

1 *According to another Native legend:* Joseph Bruchac, *Flying with the Eagle, Racing the Great Bear*, pp. 73–75.

2 *occasions of human woe:* Thornton Wilder, *The Bridge of San Luis Rey*, p. 7.

2 *thousands of little facts:* Wilder, p. 9.

3 *At least forty-four people were killed:* Telephone interview with John Jensenius, NOAA, January 15, 2008. Also see http://www.lightningsafety.noaa.gov/.

3 *Lightning is the second-leading cause:* National Weather Service, www.srh.noaa.gov/crp/stories/lightning/LightningPage.

3 *Lightning set my underclothes:* J. Todd Foster, "Story of Dooms man who survived 8 lightning strikes might play prime time," *The News Virginian*, January 3, 2003.

3 *He was first hit:* Thomas French and Sheryl James, "Lightning, Nature's Strike Force," *St. Petersburg Times*, July 23, 1989.

4 *I can be standing in a crowd:* Lee Hotz, "Lightning Strikes Man for the Seventh Time," *The News Virginian*, summer, 1977.

4 *I'm just allergic to lightning:* Lee Hotz.

4 *I actually saw the bolt:* French and James.

4 *He recalled that it was the twenty-second bear:* Lee Hotz.

5 *Naturally people avoided me:* French and James.

5 *I don't believe God is after me:* French and James.

5 *Roy Sullivan shot and killed himself:* UPI, September 29, 1983, Dooms VA.

5 *She is happily married:* Interview with Linda Cooper, June 8, 2007.

6 *Men account for about four times more: Space Science News,* NASA, June 18, 1999, citing "Demographics of U.S. Lightning Casualties and Damages from 1959–1994," Ronald L. Holle and Raúl E. López of the National Severe Storms Laboratory and E. Brian Curran of the National Weather Service.

11 *Mize was born in Sulphur, Oklahoma:* Interview with Carl Mize, May 16, 2007.

18 *God's burning finger:* Herman Melville, *Moby Dick,* p. 498.

THE MOUNTAIN CLIMBER: THE ASCENT

19 *the warm Wyoming air:* Jerry Beilinson, "Struck by Lightning on Grand Teton: Behind the Rescue," *National Geographic Adventure,* September 17, 2003, p. 1, nationalgeographic.com.

19 *smooth walls when they are dry:* Sharon Cohen, "A lightning strike in the Tetons leaves a climber dangling," Associated Press, October 18, 2003, p. 2.

22 *one in 750,000:* Ronald Holle, private correspondence with the author; the figure of one in six million is derived by dividing fifty into 300 million.

24 *most complex recovery operation:* Case Incident Record, Lightning Strike Fatality/Mountain Rescue, and Supplementary Case/Incident Record, National Park Service, United States Department of the Interior, July 26, 2003. Much of the rescue account is based on these reports.

THE ANGRY SKY

25 *The oldest representation:* Dacre Stubbs, *Prehistoric Art of Australia,* pp. 82–84.

26 *In ancient Egypt:* Vladimir A. Rakov and Martin A. Uman, *Lightning: Physics and Effects,* p. 1.

26 *The Chinese told of Lei Tsu:* Rakov and Uman, *Lightning: Physics and Effects,* p. 1.

26 *Vedic books of India described Indra:* Rakov and Uman, p. 1.

26 *some statues of Buddha show:* B. F. J. Schonland, *The Flight of Thunderbolts,* p. 3.

26 *The Yoruba:* Hal Horton, "Yoruba Religion and Myth," www.thecore.nus.edu.sg/post/nigeria/yorubarel.html.

26 *worshipped Ah-Peku: Encyclopedia Mythica,* www.pantheon.org.

26 *For the Tibetans:* http://www.brown.edu/Research /BuddhistTempleArt/ TibetanArt2.

26 *Like the Native Americans:* Schonland, pp. 1–2.

27 *an awesome flame:* Hesiod, "The Theogony of Hesiod," *The Homeric Hymns and Homerica* translated by Hugh G. Evelyn-White, p. 129.

27 *the bolt that never sleeps:* Edith Hamilton, *Mythology,* p. 83.

27 *holding the lightning:* Hesiod, p. 83.

28 *Jove let fly:* Homer, *The Odyssey,* Book XII, The Internet Classics Archive, http://classics.mit.edu/Homer /odyssey.12.xii.htm.

28 *Zeus discovers Iasion: Britannica Online Encyclopedia,* www.britannica.com/eb/article-9041837\ Iasion.

28 *Anchises:* Britannica Online Encyclopedia, www.britannica.com/eb/article-9007401/Anchises.

28 *Asclepius, the deity of medicine:* Thinkquest.org.

28 *Phaeton convinced his father:* the bakken.org.

28 *You may have observed how:* Herodotus, *Herodotus,* tr. by Reverend William Beloe, p. 325.

29 *A stomachache and then suddenly your belly:* Aristophanes, *The Clouds,* The Internet Classics Archive, classics.mit.edu /Aristophanes/clouds.html.

29 *According to Roman legend:* James G. Frazer, *The Golden Bough,* p. 172.

30 *Salmoneus, a founder of the city of Salmonia:* Frazer, p. 89.

30 *who came down:* Frazer, p. 185.

30 *who always carried a sealskin:* Suetonius, *The Twelve Caesars,* p. 103.

31 *"I will watch the sky" became a euphemism:* Philip Dray, *Stealing God's Thunder: Benjamin Franklin's Lightning Rod and the Invention of America,* p. 64.

31 *The Roman historian Suetonius:* Suetonius, *The Twelve Caesars,* pp. 310–311.

32 *there were thunders and lightnings:* Exodus, 19:16, *The Holy Bible containing the Old and New Testaments, King James Version.* Cleveland and New York: World Publishing Company, n. d.

32 *Whenever the sky wore an ugly look:* Suetonius, p. 149.

33 *There was a tradition:* Plato, *Phaedrus,* www.greektexts.com /library/Plato/phaedrus/eng/920.html.

33 *St. Boniface ordered destroyed:* www.icons.org.uk/theicons /collection/oaktree/features/the-sacred-history-of-oak-trees.

34 *Thunor was often worshipped elsewhere:* www.icons.org.uk /theicons/collection/oaktree/features/the-sacred-history-of-oak-trees.

34 *derived from the much greater frequency:* Frazer, p. 821.

34 *Shakespeare wrote:* William Shakespeare, *King Lear,* p. 1158.

34 *There was a sudden flash:* Leo Tolstoy, *Anna Karenina,* p. 917.

34 *a former Virginia state police official:* Interview with Frank Williams, June 8, 2007.

35 *the good and beautiful:* Frazer, pp. 703–704.

35 *from fire and water:* Frazer, p. 704.

35 *too young to swear:* Frazer, p. 704.

35 *flame of lightning smoldered* and *death by a stroke of lightning:* Frazer, p. 823.

35 *worshipped a mistletoe-bearing oak:* Frazer, p. 822.

36 *In some parts of Flanders and France:* Frazer, p. 738.

36 *acorn-shaped knobs:* John Matthews, "Fear of Lightning

Grounded in Myth," ASU Research, http://researchmag
.asu.edu/stories/lightning.html.

36 *Lightning poisons wine:* Dray, p. 65.

36 *Barbara Dioscorus: The Catholic Encyclopedia,* The
Encyclopedia Press, Vol. 2, 1913, p. 285.

37 *During the night of Saturday:* http://inamidst.com/lights
/stelmo.

37 *All the yard-arms were tipped:* Melville, p. 497.

38 *a number of early Greek philosophers:* Claudia Hetze,
"Lightning History," www.meteoros.de/light
/blitze_he.htm.

39 *Tertullian:* Andrew Dickson White, *A History of the Warfare
of Science with Technology,* Vol. 1, pp. 323–324.

39 *Rains and winds:* White, p. 337, quoting from the *Summa
Theologica* of St. Thomas Aquinas.

39 *It was said that the lightning strikes the sword:* White, p. 338.

40 *the sacred formula of the consecration of the mass:* Dray, p. 66,
quoting from *Appleton's Journal of Popular Literature,
Science, and Art,* October 1869.

40 *extraordinairie lightning came into the church:* Mary Reed,
"Weather Talk," *Weatherwise,* June 1988, p. 173.

40 *And in 1652:* Reed, p. 172.

40 *the thunderbolt had respect:* White, p. 332.

41 *consumed certain parts of his body:* White, p. 333.

41 *notable offenders are struck down:*
www.folger.edu/eduPrimSrcDtl.cfm?psid=137.

41 *of all the instruments of God's vengeance:* White, p. 333.

41 *Jesuit Georg Stengel claimed:* White, p. 334.

42 *Olaus Magnus:* White, p. 349.

42 *the hellish legions to flight:* White, p. 350.

42 *the surest remedy against thunder:* White, p. 350.

42 *inscriptions on medieval bells:* Dray, p. 69.

42 *"baptized" with water:* Dray, p. 67.

43 *Whensoever this bell shall sound:* White, pp. 346–347.

43 *A Proof that the Ringing of Bells:* Schonland, p. 8.

43 *baptized against lightning:* Dray, p. 68.

43 *An Israelite of the tribe:* Philippians 3:5.

44 *suddenly there shined round about him:* Acts, 9:3, 4, 8.

44 *and putting his hands on him:* Acts, 9:17, 18.

45 *I will become:* Martin Brecht, *Martin Luther: His Road to Reformation, 1483–1521*, tr. James L. Schaaf, p. 48.

STORM CHASER

47 *For forty-nine weeks:* Paul Weingarten, "The Tornado Chaser's Vacations Take a Strange Twist," *Chicago Tribune*, June 4, 1980.

47 *I don't think of myself:* Interview with David Hoadley, May 25–May 29, 2006.

48 *Part of the satisfaction of chasing:* Weingarten.

53 *When I was going to school up in Fargo, North Dakota:* Tim Marshall, "Roger Jensen: A Storm Chasing Pioneer," www.stormtrack.org/jensen.

54 *It's for the awe:* Marshall.

54 *That man:* David Hoadley, "Roger Jensen—A Friend," from the WX-Chase List, May 14, 2001, www.onthefront .ws/jensen.htm.

60 *a cloud is not simply a cloud:* Weingarten.

71 *It is not something:* David Hoadley, "Why Chase Tornadoes?" *Stormtrack*, www.stormtrack.org/library/chasing/whychas.htm.

FRANKLIN'S HERETICAL ROD

75 *All things abroad:* Jonathan Edwards, "A Faithful Narrative," Carl Van Doren, ed., *Benjamin Franklin and Jonathan Edwards*, p. 315.

75 *the temperature of the air:* Jonathan Edwards, "Things to Be Considered, or Written Fully About: Thunder," Van Doren, p. 234.

75 *an almost infinitely fine:* Van Doren, p. 234.

76 *from Isaac Newton's principles:* Jonathan Edwards, *The Works of President Edwards*, Vol. 1, p. 49.

76 *And scarce any thing:* Edwards, "Personal Narrative," Van Doren, p. 348.

76 *rejoicing in the electric shock:* Van Doren, p. xix.

76 *On the afternoon of May 10:* Joseph Priestley, *The History and Present State of Electricity with Original Experiments*, pp. 381–384.

77 *When Thomas François D'Alibard:* Dray, pp. 81–82.

77 *Gilbert had found:* Schonland, p. 15.

78 *It is not entirely clear:* Dray, pp. 37–38.

78 *electricity parties:* Dray, p. 51.

78 *Electricity is a vast country:* Quoted by Dray, p. 46.

79 *electrified clouds:* Walter Isaacson, *Benjamin Franklin: An American Life*, p. 138.

79 *the electric fluid:* Scientific minutes by Franklin from November 7, 1749, quoted in Dray, p. 57.

79 *Machine or Kite:* Benjamin Franklin, letter to Peter Collinson, 1752, quoted in Dray, pp. 89–90.

80 *The kite being raised:* Priestley, p. 217.

81 *was intrigued:* Dray, p. 88.

81 *on Sunday: Pennsylvania Gazette*, August 13, 1752, quoted in Dray, p. 87.

81 *will go considerably:* Franklin, letter to Peter Collinson, September 1753, quoted in Dray, p. 88.

82 *The Method is this: Poor Richard's Almanac*, 1753, published October 1752, reproduced in Schonland, plate II.

83 *In 1770:* Peter E. Viemeister, *The Lightning Book: The Nature of Lightning and How to Protect Yourself From It*, p. 47.

83 *the Westinghouse Company:* Viemeister, p. 48.

84 *as impious:* Dray, p. 96.

84 *It has pleased God: Poor Richard's Almanac*, Schonland, plate II.

84 *Thunder of Heaven:* Franklin, letter to Cadwallader Colden,

April 12, 1753, quoted in I. Bernard Cohen, *Benjamin Franklin's Science*, p. 141.

84 *not so extraordinary:* Franklin, letter to John Winthrop, July 2, 1768, quoted in Dray, pp. 100–101.

85 *all these iron points:* Schonland, pp. 28–29.

85 *seemed to float:* Dray, p. 97.

85 *globe of blue fire:* Priestley, p. 418.

86 *an inconsiderable quantity of blood:* Priestley, 419.

86 *distressing season:* Dray, p. 158.

86 *an old quibbler:* Jessica Riskin, *Science in the Age of Sensibility: The Sentimental Empiricists of the French Enlightenment*, p. 149.

86 *a sword:* Riskin, p. 139.

87 *uncanny occurrence:* Dray, p. 154.

87 *no miracle here:* Dray, p. 160.

88 *The more points:* Dray, p. 105.

88 *Strange as it may seem:* Cohen, p. 146.

88 *our duty:* Brooke Hindle, *The Pursuit of Science in Revolutionary America, 1735–1789*, pp. 95–96.

89 *This invention:* Zoltan Haraszti, "Young John Adams on Franklin's Iron Points," *Isis*, Vol. 41, No. 1, p. 13.

89 *by micromanaging*: Isaacson, p. 86.

89 *doing good to man*: Isaacson, p. 84.

89 *I imagine it great vanity*: Isaacson, p. 85.

89 *divine wrath:* Cohen, p. 142.

91 *no private interest:* Schonland, p. 28.

92 *Between 1799 and 1815:* Frank W. Lane, *The Violent Earth*, p. 92.

92 *the oak draws lightning:* Herman Melville, "The Lightning-Rod Man," www.classicshorts.com/stories/tlrm.htm.

94 *a young woman is presenting:* For connections to lightning in Niquet le Jeune's gravure and other works discussed, I am indebted to Christian Fuhrmeister and to his "The Political Iconography of Lightning," presented at Taming the Electrical Fire: A Conference on the History and Cultural

Meaning of the Lightning Rod, The Bakken Library and
Museum, Minneapolis MN, November 4–6, 2002.

Isamu Noguchi honored Franklin with his *Bolt of
Lightning* sculpture in Philadelphia. Other cultural works
involving lightning include Walter Gropius's *Monument to
the March Dead*, in Weimar, Germany, which depicts
lightning rising from workers' graves (it was attacked
beginning in the 1930s by the Nazis, who used the symbol
of double lightning for their own purposes); Francis Ford
Coppola's film *Youth Without Youth*, based on the novel by
Mircea Eliade; Dean Koontz's novel *Lightning*; and
Jonathan Borofsky's 70-foot-high painted steel sculpture
Lightning in Tampa, Florida.

95 *At the moment:* White, p. 364.

THE ODD COUPLE

98 *Nothing is really known:* Interview with Martin Uman,
January 7, 2007.

98 *who also teaches at the University of Florida:* Interview with
Vladimir Rakov, January 7, 2007.

101 *estimated 1.2 billion lightning flashes:* "Space Research and
Observations," NASA, http://thunder.msfc.nasa.gov/otd.
The estimate is based on NASA's Optical Transient
Detector (OTD), a compact combination of optical and
electronic elements. The name, Optical Transient
Detector, refers to its capability to detect the momentary
changes in an optical scene that indicate the occurrence of
lightning. The OTD instrument is a major advance over
previous technology as it can gather lightning data under
daytime conditions as well as at night.

101 *more than ten thousand fires:* Richard Kithil, "Annual USA
Lightning Costs and Losses," National Lightning Safety
Institute, http://www.lightningsafety.com/nlsi_lls/
nlsi_annual_usa_losses.htm. In 2006, there were about
16,000 lightning-caused fires in the U.S., according to the

National Interagency Fire Center, "Fire Information—Wildland Fire Statistics," http://www.nifc.gov/fire_info/lightning_human_fires.html.

101 *$5 billion in annual damages:* Richard Kithil, "Annual USA Lightning Costs and Losses," National Lightning Safety Institute, http://www.lightningsafety.com/nlsi_lls/nlsi_annual_usa_losses.htm.

102 *to as much as 50,000 degrees Fahrenheit:* Martin Uman, *All About Lightning,* p. 93.

103 *Tall objects:* For a fuller description, see Viemeister, pp. 112–113.

103 *Twisting and turning:* Viemeister, p. 113.

105 *Tesla alerted his mechanic:* "Tesla Life and Legacy: Colorado Springs," PBS, http://www.pbs.org/tesla/ll/ll_colspr.html.

106 *General Electric:* Floyd Miller, *The Man Who Tamed Lightning: Charles Proteus Steinmetz,* pp. 47–48.

107 *Wonderful:* Miller, p. 142.

107 *hundred-thousandth of a second:* Sigmund A. Lavine, *Steinmetz: Maker of Lightning,* p. 212.

107 *the terror:* John Winthrop Hammond, *Charles Proteus Steinmetz: A Biography,* p. 454.

108 *scientific foundation of religion:* Hammond, p. 459.

108 *Frankenstein-like:* Miller, pp. 145–146.

108 *football goalpost:* Sigmund A. Lavine, *Steinmetz: Maker of Lightning,* p. 210.

109 *modern Jove:* *The New York Times,* March 2, 1922.

109 *Lightning has always been:* Hammond, pp. 330–331.

109 *The pint of gasoline:* Hammond, p. 338.

111 *progresses downward:* Brian Austin, *Schonland: Scientist and Soldier,* p. 119.

112 *a blob of electrons:* Viemeister, pp. 110–111.

112 *like a snake:* Austin, p. 153.

113 *This discovery:* Austin, p. 156.

115 *they're beautiful:* See Fred W. Wright, Jr., "Florida's

Fantastic Fulgurite Find," *Weatherwise*, July/August 1998, p. 29.

115 *Neiman Marcus:* Brian Conley and Sina Najafi, "All About Lightning: An Interview with Martin Uman," *Cabinet*, Issue 3, Summer 2001, www.cabinetmagazine.org/issues/3/allaboutlightning.php.

116 *a high-tech version:* Robert Roy Britt, "Zap! Rockets Trigger Lightning, Scientists Discover X-Rays," www.space.com/businesstechnology/technology/rocket_lightning_030130.html.

116 *Triggering lightning:* Uman and Rakov, p. 265.

117 *big objects:* Conley and Najafi, p. 2.

119 *about eighty feet from the tower:* Britt.

119 *eighty-seven percent:* "Florida Researchers: Lightning Emits X-Rays; Modern-Day Ben Franklins Use Rockets to Settle 80-Year-Old Debate," University of Florida News, January 30, 2003, http://news.ufl.edu/2003/01/30/lightxray/.

119 *visible stroke:* www.scienceblog.com/ems/flash_lightning _emits_x-rays.

123 *strong electromagnetic field:* Medley O'Keefe Gatewood and Richard Zane, "Lightning Injuries," *Emergency Medicine Clinics of North America*, 22, 2004, p. 371.

126 *In 1936:* Viemeister, p. 130.

126 *He was upstairs:* Viemeister, p. 131.

126 *another remarkable account:* Interview with Jean Zaleski, February 15, 2007.

127 *no adequate theory:* Martin Uman, *All About Lightning*, p. 131. "Human beings are seldom, if ever, killed by ball lightning," according to Uman and Rakov, *Lightning: Physics and Effects*, p. 656.

127 *a small percentage of the UFO reports:* Uman, *All About Lightning*, p. 133.

128 *energies from sprites:* "Spirits of Another Sort," NASA Science News, http://science.nasa.gov/newhome/headlines/essd10jun99_1.htm.

THE MOUNTAIN CLIMBER: THE SUMMIT

133 *crafted from a block:* John C. Reed, Jr., "Geology of the Teton Range" in Leigh N. Ortenburger and Reynold G. Jackson's *A Climber's Guide to the Teton Range,* p. 39. The following section on the history of the Tetons relied primarily on this guide.

138 *burned horrifically:* Beilinson, p. 2.

140 *The guy's alive:* Sharon Cohen, "A lightning strike in the Tetons leaves a climber dangling," p. 6.

140 *That guy's gonna die:* Cohen, "A lightning strike in the Tetons," p. 6.

141 *Bloody hell:* Cohen, "Helicopter fights winds and rangers scramble," p. 2.

142 *We're going to have to abort:* Cohen, "Helicopter," p. 2.

143 *We've got to get people:* Cohen, "Helicopter," p. 2.

143 *You've got to catch Springer:* Cohen, "Helicopter," p. 2.

144 *flat rock:* Cohen, "Helicopter," p. 3.

144 *a wee man:* Cohen, "Helicopter," p. 3.

145 *Is he alive?:* Cohen, "Helicopter," p. 3.

145 *Hang in there:* Cohen, "Helicopter," p. 4.

148 *Get ready for the best ride:* Cohen, "Helicopter," p. 4.

150 *I'm going to need an extra hand:* Cohen, " 'Am I going to be OK?' dangling climber asks as rangers finally reach him," p. 2.

151 *Let's go:* Cohen, " 'Am I going to be OK?' " p. 4.

UNEXPECTED CONSEQUENCES

154 *golfing accident:* "Lightning strike hurts 19 during Colorado golf event," USA Today.com, 6/20/2004, http://www.usatoday.com/news/nation/ 2004-06-20-lightning-colorado_x.htm.

154 *Sam Snead:* http://golf.about.com/od/golfersmen/a/snead_quotes.htm.

154 *Lee Trevino:* thinkexist.com/quotes/lee_trevino/.

154 *24,000 people are killed:* Ronald L. Holle and Rául E. López,

"A Comparison of Current Lightning Death Rates in the U.S. with Other Locations and Times," 2003, *Preprints, International Conference on Lightning and Static Electricity*, Blackpool, England, 2003.

155 *In the developed countries:* Ronald L. Holle, "Annual Rates of Lightning Fatalities by Country," paper presented at the International Conference on Lightning and Static Electricity, Paris, France, 2007.

155 *in Germany, three to seven:* Manfred Dworschak, "Lightning Strike Survivors Meet for World Conference," Part 3, Spiegel Online International, www.spiegel.de/international/zeitgeist/0,1518,491477,00.html.

155 *In the U.S.:* Ronald Holle, private correspondence with the author, September 6, 2007.

155 *a hundred annual deaths:* E. B. Curran, R. L. Holle, and R. E. López, "Lightning casualties and damages in the United States from 1959 to 1994," *Journal of Climate*, 13, 2000, pp. 3448–3453. Data is also available at the National Severe Storms Laboratory Web site: http://www.nssl.noaa.gov/papers/techmemosNWS-SR-193/techmemo-sr193.html.

155 *350 and 550 injuries:* National Weather Service, Colorado Lightning Resource Page, http://www.crh.noaa.gov/pub/?n=ltg.php, and Ronald Holle, private correspondence with the author, September 6, 2007. There were sixty estimated deaths in 2006.

155 *anecdotal evidence:* National Weather Service, "Updated AMS Recommendations for Lightning Safety—2002," www.lightningsafety.noaa.gov.

156 *Dave Grillmeier:* "Is It Worth One More Cast?," http://www.ohiogamefishing.com/community/archive/index.php/t-13346.html, and telephone interview with Dave Grillmeier, November 18, 2007.

156 *Neponsit Beach:* NYC Office of Emergency Management, www.nyc.gov/html/oem/hazards/weather_thunder.shtml.

157 *William Snow Harris:* Dray, p. 181.

157 *small boats are seldom made:* William J. Becker, "Boating-Lightning Protection," www.cdc.gov/nasd/docs/d000001-d000100/d000007/d000007.html.

157 *a man and his nine-year-old son:* Struckbylightning.org, p. 101 of database.

158 *1995 in Honduras:* Ronald Holle, private correspondence with author, September 6, 2007.

158 *A prominent American accident:* Stephen J. Dollinger, "Lightning-strike disaster among children," *British Journal of Medical Psychology*, 58, 1985, p. 375.

158 *a high-school player died:* Christopher Lawlor, "Prep player dies after lightning strike," *USA Today* (September 16, 2004).

159 *official Scout Jamboree manual:* "Philmont Health and Safety," www.scouting.org/philmont/camping/tips/safety.html.

159 *Mark Evans:* Boy Scouts of America, 2001 National Scout Jamboree, p. 2, www.scouting.org/media/reports/2001/jamboree.html.

159 *Mark and his group:* Telephone interview with Mark Evans, July 8, 2007.

160 *seven scouts and scout leaders:* David Crary, "Public Safety, Are Scouts Put at Risk," Associated Press, January 15, 2006.

160 *William Roeder:* Crary.

160 *Matthew Tresca:* Crary.

161 *I saw a lot of people:* Wayne Parry, "Official: Scouts sent campers to tents during storm," phillyBurbs.com.

161 *Another camp employee:* Parry.

161 *Scout officials denied:* Crary.

162 *If only on the basis:* Crary.

162 *Mary Tresca:* Crary.

162 *James Rozwood:* Crary.

162 *a number of scouts:* Vernon Loeb, "A Killer Bolt out of the Blue," *Los Angeles Times*, August 16, 2005.

163 *My recommendation:* Loeb.

163 *We take lightning:* Gregg Shields, private communication with author, July 12, 2007.

164 *He suffered burns:* Linda Johnson, "Lightning strikes reported by iPod users," Associated Press, July 11, 2007.

165 *A thirty-seven-year-old jogger:* Lester Haines, "Canadian iPod user struck by lightning," *The Register*, http://www.theregister.co.uk/2007/07/13/ipod_lightning_strike.

165 *There is no evidence:* Chris Andrews and Mary Ann Cooper, "To the Editor," *The New England Journal of Medicine*, 357, 14, p. 1447.

165 *basic safety rules:* "Updated AMY Recommendations for Lightning Safety—2002," National Weather Service.

MEDICAL DETECTIVES

167 *fatal injury to only about nine to ten percent:* Mary Ann Cooper, "Disability, Not Death, Is the Main Problem with Lightning Injury," Abstract from an address presented at the National Weather Association annual meeting, Oklahoma City, Oklahoma, October 1998, http://www.uic.edu/labs/lightninginjury/Disability.pdf.

167 *It hit the pole:* Interview with Tony Scott, January 13, 2007.

168 *The doctor there:* Brad and Linda Anderson, *Life After Shock*, Vol. III, pp. 120–121.

169 *There was a strange hissing sound:* Interview with Lisa Hall (formerly Cunningham), June 8, 2007, and Lisa Cunningham, "The Mystery Condition," *Life After Shock*, Vol. II.

169 *a marathon:* Interview with Paul Aurand, April 3, 2007.

170 *vast constellation:* Richard F. Edlich and David B. Drake, "Burns, Lightning Injuries," eMedicine, eMedicine.com/plastic/topic517.htm.

171 *100 million to a billion volts:* "Interesting Facts, Myths, Trivia and General Information about Lightning," National Weather Service Forecast Office, Melbourne, Florida, http://www.srh.noaa.gov/mlb/ltgcenter /ltg_facts.html.

173 *most sadly:* Charles "Nick" Simkins, "Can Long-Term Cognitive and Emotional Problems Be Caused by Electric Shock and Lightning Strike Accidents or Is Anything That I Know About Brain Injury Applicable to Electric Shock and Lightning Strike Victims?," paper written for Lightning Strike and Electric Shock Survivors International, n.d., www.lightning-strike.org/ DesktopDefault.aspx?tabid=63.

173 *When a survivor walks into a doctor's office:* Interview with Dr. Mary Ann Cooper, May 24, 2007.

176 *by-now classic 1980 article:* Mary Ann Cooper, "Lightning Injuries: Prognostic Signs for Death," *Annals of Emergency Medicine,* 9:3, March 1980.

176 *No inconvenience:* Priestley, pp. 174–175.

176 *I never knew any advantage:* Franklin, letter to John Pringle, Dec. 21, 1757, www.historycarper.com/resources/ twobf3/letter1.

177 *a British physician:* Michael Cherington, "James Parkinson; Links to Charcot, Lichtenbergh, and Lightning," *Archives of Neurology,* Vol. 61, June 2004, np.

177 *Jean-Martin Charcot:* Christopher J. Andrews, Mary Ann Cooper, Mat Darveniza, David Mackerras, *Lightning Injuries: Electrical, Medical, and Legal Aspects,* p. 88.

177 *As soon as the patient becomes aware:* Macdonald Critchley, "Neurological Effects of Lightning and of Electricity," *The Lancet,* January 13, 1934, p. 71.

178 *Most victims of lightning:* Orthello R. Langworthy, "Neurological abnormalities produced by electricity," *Journal of Mental Disorders,* 84, 1936, pp. 18–26.

178 *none of the patients:* Thomas Muehlberger, Peter M. Vogt,

Andrew M. Munster, "The long-term consequences of lightning injuries," abstract, *Burns,* 27, 2001, p. 829.

178 *Many physicians:* Simkins, p. 1.

179 *over a thousand times:* H. Wachtel, Michael Cherington, and Philip Yarnell, "Sub-Visible, Low Thermal, Lightning May Account for Cases of 'Unexplained' Sudden Cardiac Death," presented at the Annual Conference of the Bioelectromagnetics Society, Cancún, Mexico, June 2006.

181 *You may also find:* Gene Madden, "The Dos & Don'ts of Lightning Safety," *Wildland Firefighters* magazine, Vol. 24, Issue 6, June 2006, http://www.firerescue1.com/ wildland-firefighter/24-6/106548/.

182 *another leading medical authority:* The small group of medical and psychiatric authorities on lightning includes, besides those mentioned in the text, Neil H. Pliskin, Ph.D.; Dr. Nelson Hendler; Gerolf H. Engelstatter, Ph.D.; Margaret Primeau, Ph.D.; and Dr. Ryan Blumenthal.

182 *exaggerates "on the other side":* Mary Ann Cooper, private correspondence with the author.

183 *Some of the folks:* Harold R. Deal, "A Truly Unforgettable and Thankful Story," *Life After Shock,* Vol. I, pp. 15–16.

184 *A year before:* Interview with Elizabeth (Betty) Galvano, January 12, 2007.

185 *as though his scar were on fire:* J. K. Rowling, *Harry Potter and the Sorcerer's Stone,* p. 256.

185 *a warning . . . it means:* Rowling, p. 264.

186 *Charles Xavier:* www.marvel.com/universe/Professor_X.

186 *supersensitive to electrical current:* "Gabrielle Blanstein," *Life After Shock,* Vol. II, pp. 44–45.

187 *They come back on:* "Nina Lazzeroni," *Life After Shock,* Vol. II, pp. 152–153.

187 *I looked like Felix the Cat:* Interview with Kurt Oppelt, January 10, 2007.

188 *Lightning Data Center:* Interviews with Michael
    Cherington, Philip Yarnell, and Howard Wachtel,
    November 10, 2006.
188 *struck at the industrial plant:* "Jerry D. and Bee deDoux
    [*sic*]," *Life After Shock*, Vol. III, p. 73.
188 *My lower-back pain:* Antoinette (Toni) Palmisano,
    "Stiffened by a Lightning Bolt," *Life After Shock*, Vol. III,
    p. 17.
189 *It's a pain that makes you feel:* Interview with Steve
    Marshburn, Sr., June 9, 2007.

A GATHERING OF ANGELS

192 *There are times:* Interview with Ron "Griz" Swidorsky,
    June 9, 2007.
194 *I thought of those humans:* Gretel Ehrlich, *A Match to the
    Heart*, p. 160.
195 *I was too young:* Judith Friend, *Life After Shock*, Vol. III,
    pp. 40–41.
196 *one seems to be:* Oliver Sacks, "A Bolt from the Blue," *The
    New Yorker*, July 23, 2007, p. 41.
196 *are not easily dismissed:* Sacks, p. 42.
197 *We crawled over to her:* Interview with Erin, Elizabeth, and
    Edward Ellickson, June 7, 2007.
197 *near-death experience:* Susan Blackmore, *Dying to Live: Near
    Death Experiences*, p. 4.
198 *neural correlates:* Sacks, p. 41.
198 *It sounded like a bomb:* Interview with Lindsey Sass-Aurand,
    April 3, 2007.
204 *We get a lot of lightning:* Interview with Jeffrey Bryer,
    January 13, 2007.
207 *old superstitious lady:* Interview with Veronica Smith, April
    7, 2007.
208 *I have angels everywhere:* Interview with Nathan, Becky, and
    Paul Loyet, May 22, 2007.

210 *It was a black bird:* Telephone interview with Julie Spotted Eagle Horse, May 29, 2007.

213 *We loved the river:* Interview with Judy and Robert Teson and Karen Henkel Melton, August 8, 2006.

218 *I was just a typical kid:* Interview with Robert Davidson, June 8, 2007.

218 *He had no pulse:* Bobette Bryan, "The Lady in Black: A Miracle?," *Underworld Tales Magazine,* 2005, www.underworldtales.com/miracle.

219 *There is no doubt:* Bryan.

228 *simple-minded ol' farm boy:* Interview with Gwynn Blair, January 9, 2007.

THE MOUNTAIN CLIMBER: THE DESCENT

231 *Clipped and ready:* Cohen, " 'Am I going to be OK?' " p. 4.

231 *We're flying:* Cohen, " 'Am I going to be OK?' " p. 4.

232 *familiar landmarks:* Beilinson, p. 2.

233 *likely died:* Beilinson, p. 2.

235 *These things happen for a reason:* Cohen, " 'Am I going to be OK?' " p. 5.

EPILOGUE

242 *Braced and in place:* Richard McCord, " 'First It Reached Out On an Intellectual Plane,' " *The Santa Fe Reporter,* Vol. 5, No. 38, March 22, 1979, p. 4.

243 *When I was hit:* Peggy Noonan, "A Great Moment in the Life of an Artist: Talking with James Caviezel After His Meeting with the Pope," *Opinion Journal,* March 18, 2004, WSJ.com.

INDEX

About the Author

The producer of the Oscar-winning documentary *Hotel Terminus*, John S. Friedman has written for the *New York Times* and other publications, and contributes regularly to the *Nation*. The editor of *The Secret Histories*, he lives in Connecticut.